Subduction of Aseismic Oceanic Ridges: Effects on Shape, Seismicity, and Other Characteristics of Consuming Plate Boundaries

P. R. Vogt
U.S. Naval Oceanographic Office, Washington, D.C. 20373
A. Lowrie
U.S. Naval Oceanographic Office, Washington, D.C. 20373
D. R. Bracey
U.S. Naval Oceanographic Office, Washington, D.C. 20373
R. N. Hey
University of Texas Marine Science Institute, Galveston, Texas 77550

THE
...OGICAL SOCIETY
...ICA

Library of Congress Catalog Card Number 75-40900
I.S.B.N. 0-8137-2172-5

The Special Paper series was originally made possible
through the bequest of
Richard Alexander Fullerton Penrose, Jr.

Published by
THE GEOLOGICAL SOCIETY OF AMERICA, INC.
3300 Penrose Place
Boulder, Colorado 80301

Printed in The United Stat

Contents

Abstract

Aseismic ridges on underthrusting oceanic plates often trend into cusps or irregular indentations in the trace of the subduction zone. For example, the Hawaii-Emperor Ridge trends into the Kuril-Aleutian cusp, and the Marianas arc is bounded by the Marcus-Necker Ridge on the north and the Caroline Ridge on the south. The association between ridges and cusps is too common to be due to chance; it is proposed that the extra buoyancy of the plate with its aseismic ridge gives the plate greater resistance to sinking. This would inhibit back-arc extension and thereby produce a notch in the subduction zone. Island arcs may, therefore, acquire their curvature by additional constraints than the Earth's curvature. The geology of about 15 such cusp areas is examined for evidence to test the hypothesis that cusps were caused by subducted aseismic ridges. This hypothesis applies only to cases where extensional basins lie behind the arcs. There also appear to be cases where the trace of the subduction zone has been modified not by inhibited back-arc spreading but by splintering of the overthrusting and possibly the underthrusting plate as well. Extremely high, massive aseismic ridges might induce arc polarity reversals and thereby assume the role of protocontinental nuclei.

Seismicity and volcanism are examined where aseismic ridges are being subducted; there are several examples of reduced seismicity that cannot be explained by insufficient sampling time. By modifying the geometry of the subduction zone, the downgoing ridges necessarily affect seismicity. In addition, the plate containing the ridge may be thinner and hotter and more likely to deform by creep. There is no systematic increase or decrease in the number of andesite volcanoes where the ridges are subducted. However, lines of volcanoes and sometimes other kinds of geologic and seismic provinces may stop or start at the arc-ridge intersections. This is attributed to segmenting of the lithosphere into distinct tongues, each tongue acting more or less independently. Aseismic ridges would act as lines of weakness along which the downthrust slab becomes detached. *Key words: marine geology, marine geophysics, island arcs, subduction, plate tectonics, seismicity.*

Introduction

Subduction zones resemble a scalloped alternation of arcs and cusps or cusplike irregularities. In this paper we develop the proposal of Vogt (1973a) that many of the complexities of consuming plate boundaries are caused by the relative buoyancy of aseismic ridges on the downgoing plate (Figs. 1, 2, 3). These ridges would resist being subducted; therefore, they would also preferentially inhibit back-arc or inter-arc extension of the type proposed by Karig (1971a, 1971b, 1972) and now reasonably well established, at least for the younger marginal basins of the western Pacific. We also examine the seismicity, volcanism, and morphology of the arc-trench gap to determine whether these features of subduction zones have been affected in a systematic way where ridges are being subducted. Our discussion does not in general depend on the mode of origin of the aseismic ridges, whether they are, for example, hot-spot traces (Wilson, 1965; Morgan, 1971, 1972, 1973), fracture ridges (Menard and Chase, 1970), or remnant arcs (Karig, 1972), nor does our discussion depend on the physical processes responsible for the back-arc extension (Sleep and Toksöz, 1971).

In order to examine the phenomenon worldwide, it is necessary to limit the discussion of the numerous individual areas to keep the text to manageable proportions. For some areas, the reader must consult the cited references to evaluate the evidence cited in this text. We have emphasized regional papers, recent reviews, and critical items of evidence.

The first attempt to fit the long-recognized curvature of consuming plate boundaries into the framework of plate tectonics was that of Frank (1968) (see also Bott, 1971; and Le Pichon and others, 1973). According to Frank's hypothesis, the lithosphere is assumed to be a flexible but inextensible spherical shell whose thickness is negligible compared to the radius of the sphere. If such a shell is bent inward, the bent portion is part of an intersecting spherical surface that, because the shell is inextensible, has the same radius as the sphere. The bent portion of the shell represents the subducted tongue of lithosphere, whose definition largely depends on the so-called Benioff zone of seismicity. Although the inextensible shell has the same radius of curvature after it is subducted, it may be concave upward or downward; that is, toward the center of the Earth (Frank, 1968; Bott, 1971; Le Pichon and others, 1973). The dip of the subducted slab is the angle (α) of intersection between the bent shell and the surface of the sphere. The intersesction of the bent shell with the spherical Earth traces out a circular arc on the Earth's surface. This arc is defined by a small circle of radius $\alpha/2$ or $(\pi-\alpha)/2$, depending on whether the concavity of the bent part of

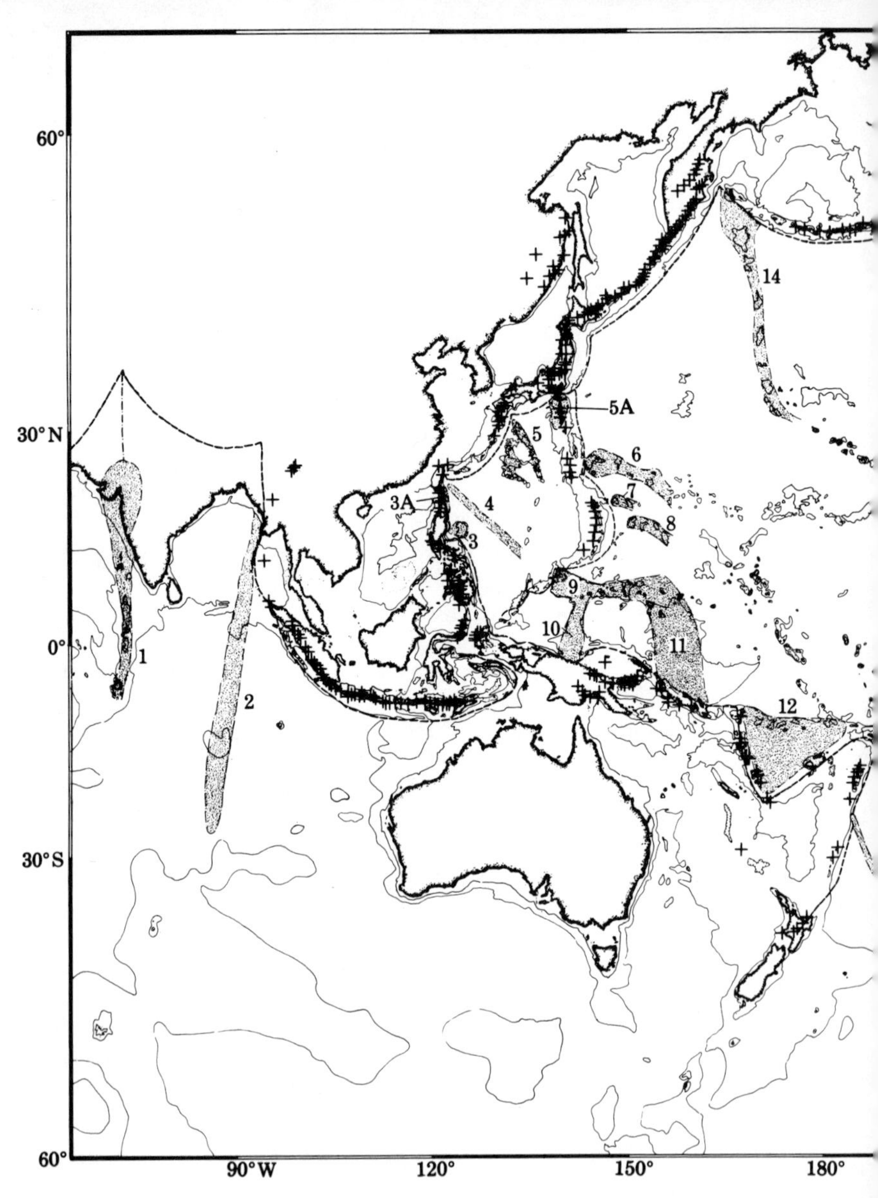

Figure 1. Aseismic ridges, platforms, or seamount chains (stippled) occurring on oceanic lithosphere in the vicinity of subduction zones. Dashed line indicates relevant plate boundaries, in most cases the axes of deep sea trenches. Crosses mark Pliocene to Holocene volcanoes or groups of close-spaced volcanoes (Anonymous, 1970a). The following ridges are indicated by number; an asterisk indicates features shown in more detail elsewhere in this paper: (1) Chagos-Maldive-Laccadive Ridge and its possible extension into continental India, in the form of the Deccan Traps; (2) Ninetyeast Ridge; (3) Benham Rise, also called Daito-Luzon Rise (Uyeda and Miyashiro, 1974); (3A) Palaui Ridge (Karig, 1973); (4) Central Basin fault or Philippine Ridge; (5) complex of ridges,

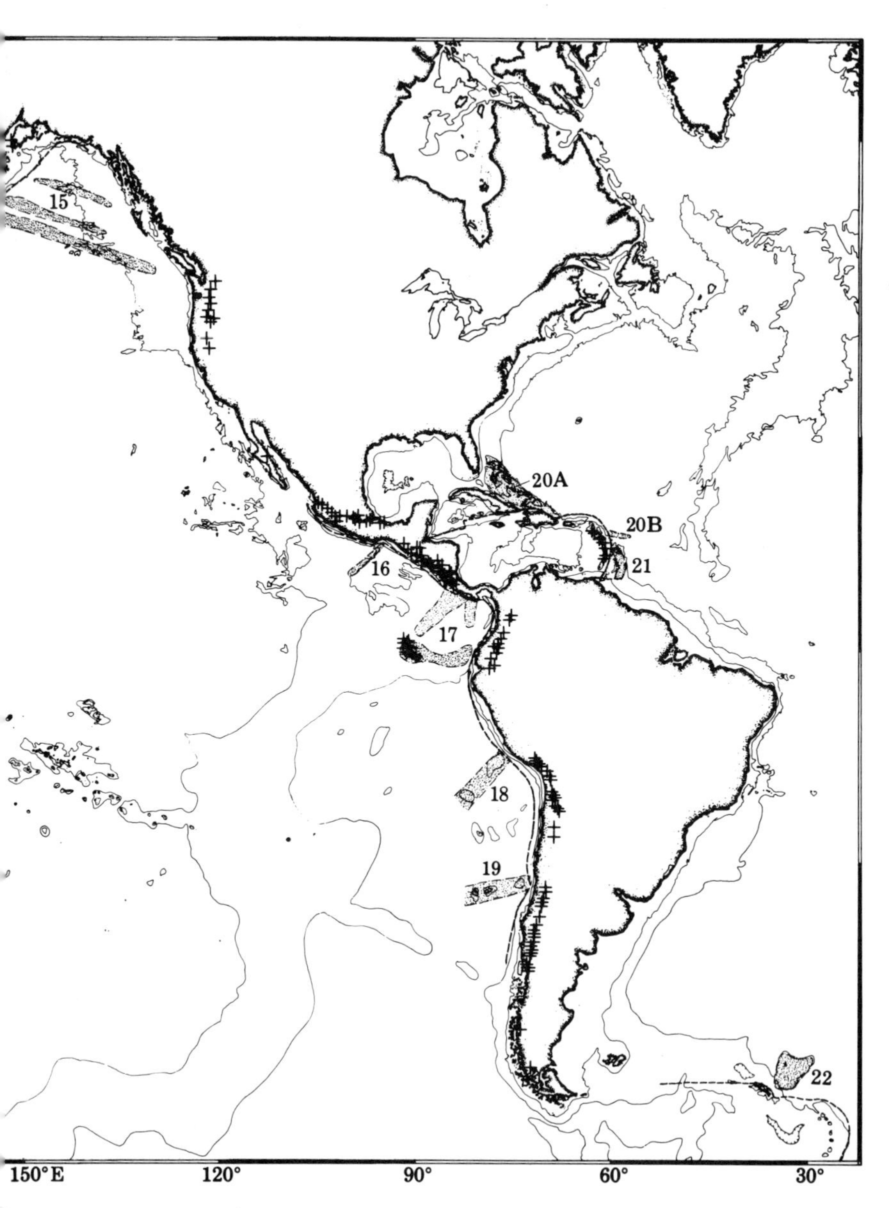

'rom north to south the Kyushu-Palau, Amami Plateau, Daito, and Oki-Daito; 5A) Shichito-Iwojima and Nishi Shichito Ridges (Uyeda and Miyashiro, 1974); 6*) Marcus Ridge (Mid-Pacific Mountains); (7*) and (8*) Magellan Seamounts; 9*) Caroline Ridge; (10) Eauripik Rise; (11) Ontong-Java Plateau; (12) Fiji Plateau; (13*) Louisville Ridge; (14*) Emperor Seamounts (Hawaii-Emperor Ridge); (15*) Pratt-Welker and Kodiak seamount chains, collectively termed Alaska Seamounts; (16) Tehuantepec Ridge; (17*) Cocos, Coiba, Malpelo, and Carnegie Ridges; (18) Nazca Ridge; (19) Juan Fernandez islands and seamounts, also called Chile fracture zone (Sillitoe, 1974); (20A*) Bahama platform; (20B*) Barracuda Ridge; (21*) Barbados Ridge; (22*) South Georgia Rise.

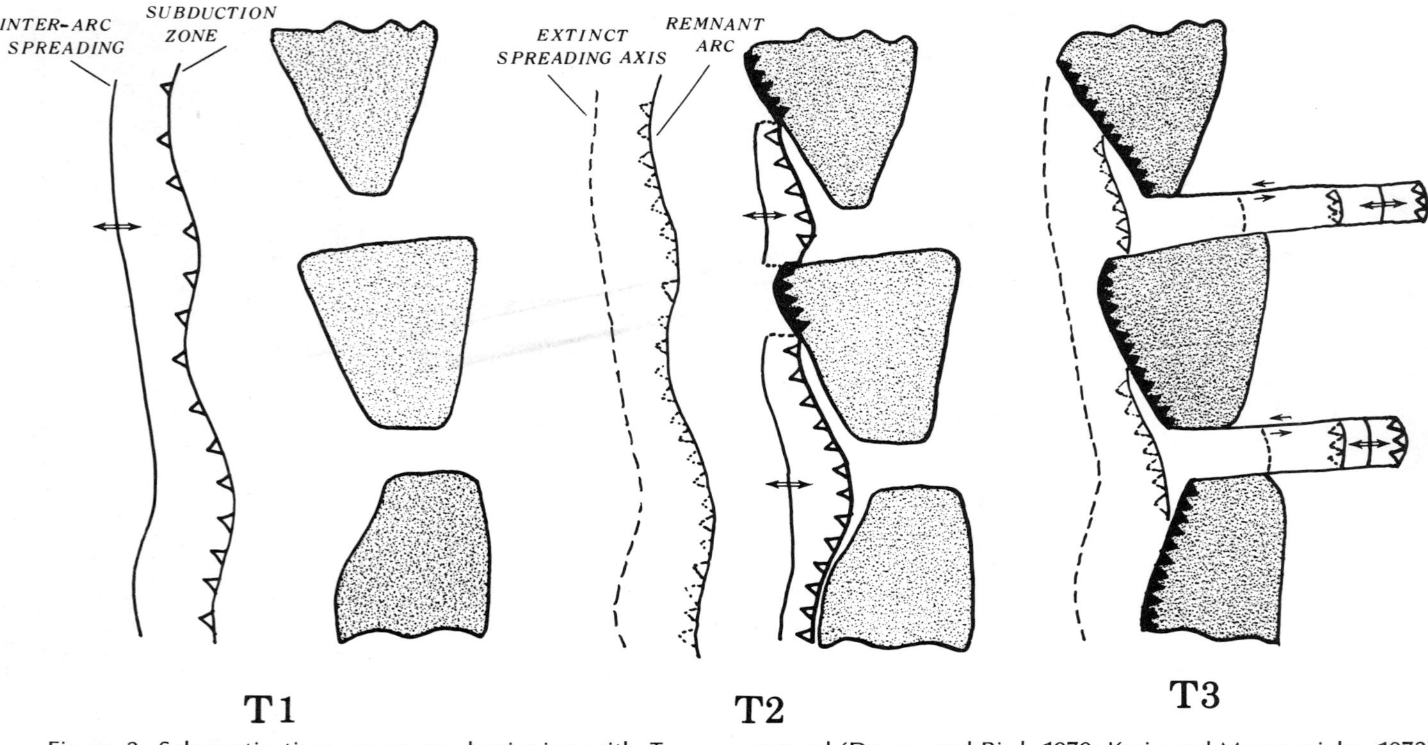

Figure 2. Schematic time sequence, beginning with T_1, illustrating eastward advance of an island arc subduction zone, behind which episodic back-arc or inter-arc spreading (Karig, 1972) occurs. Teeth show direction of underthrusting; dashed or dotted features are inactive. When arc reaches reversal (Dewey and Bird, 1970; Karig and Mammerickx, 1972) is forced to occur; inter-arc basins and ridges are now subducted under "continents" along Andean-type orogens (black teeth). Note that lithosphere is here considered to be either subductible or nonsubductible. A suggested modifi-

the shell is directed upward or downward toward the center of the Earth. In the latter case, apparently first recognized by Le Pichon and others (1973), the surface trace is concave toward the *overridden* (subducted) plate. In nature the surface traces of subduction zones are almost always concave toward the *overriding* plate. Frank's (1968) hypothesis predicts a specific relationship between the dip of the Benioff zone and the radius of curvature of the arc, and can be tested, at least in principle, by defining the exact shape of the dipping plate from the seismic activity inside the plate.

Although the inextensible bent-shell model of Frank (1968) and Le Pichon and others (1973) may explain some of the arcs, it fails on others. Furthermore, the dip of the shallow part of the seismic zone is only poorly known in most cases, so that the model cannot be put to a precise test. Where there is a gradual bending of the plate to the 45° or greater dip typical of the deeper parts of the zone (Sykes and others, 1970; Mitronovas and Isacks, 1971), the choice of what is the shallow part of the zone (Le Pichon and others, 1973) is to some extent arbitrary. Also, the assumption that the lithosphere behaves inextensibly is highly dubious; evidence suggests that plates develop permanent strain (Le Pichon and others, 1973). Even if the bent-shell model explains the curvatures of consuming plate margins to first order, we may ask why the arcs developed when and where they did—what process determined the location of individual arcs and the cusps between them? Because many of the cusps and other irregularities in the consuming plate margins are associated with ridges on the downgoing oceanic plate (Fig. 1), Vogt (1973a) suggested that these ridges, because of the extra buoyancy imparted to the oceanic plate in their vicinity, are not as readily subducted as is normal oceanic plate nearby. As a result, back-arc (also called inter-arc) extension (Karig, 1971a, 1971b, 1972; Karig and Mammerickx, 1972) would be inhibited and cusps would form where such ridges are or have recently been subducted (Fig. 3). Extensional basins behind island arcs have apparently not developed continuously, but in pulses a few million years in duration. Episodic island-arc volcanism of 10 to 20 m.y. duration (Dickinson, 1973) may be registering these pulses of activity. Accordingly, we should expect the cusps to have developed largely during such pulses of back-arc extension. Arc-shaped subduction zones would develop between the cusps, which means that the curvature of consuming boundaries is not necessarily due to the spherical shape of the Earth as postulated by Frank (1968) and elaborated by Le Pichon and others (1973). It is most probable, however, that both constraints are at work: Ridges on downgoing plates would "pin" the subduction zones and cause cusps to develop; the downthrust lithosphere between these cusps then attempts to adjust its dip and curvature to minimize strain, in order to conform to Frank's (1968) model of a bent, inextensible shell.

The mechanism we propose (Fig. 3) is difficult to prove directly because the most critical evidence would have been subducted. (However, seismicity may show whether extensions of the ridges can still be recognized in the subducted slab.) The following necessary, if not sufficient, conditions must be met, however, and we shall examine the available evidence in this light: A massive aseismic ridge intersecting a subduction zone should generally trend into a cusp, provided back-arc extension has recently occurred or is occurring. (There may be excep-

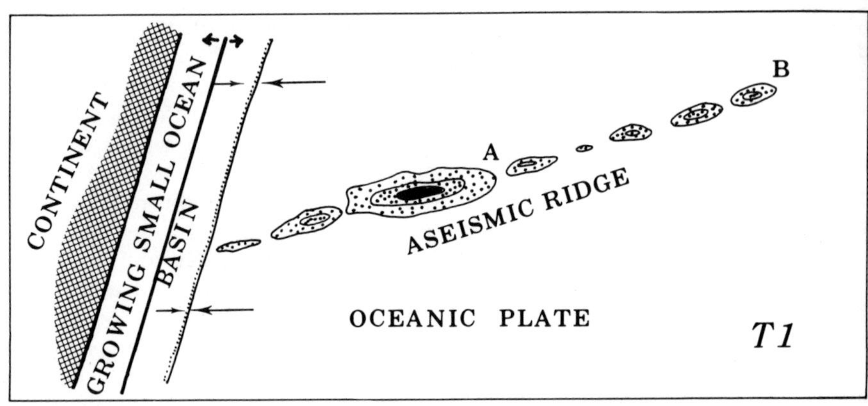
CONTINENT
GROWING SMALL OCEAN
BASIN
A
B
ASEISMIC RIDGE
OCEANIC PLATE
T1

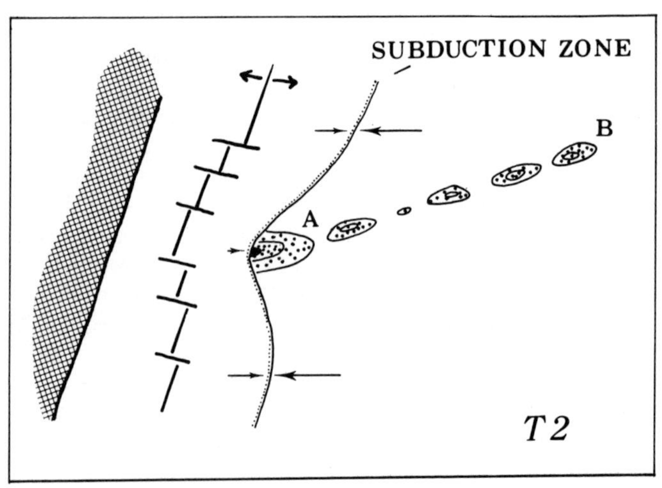
SUBDUCTION ZONE
A
B
T2

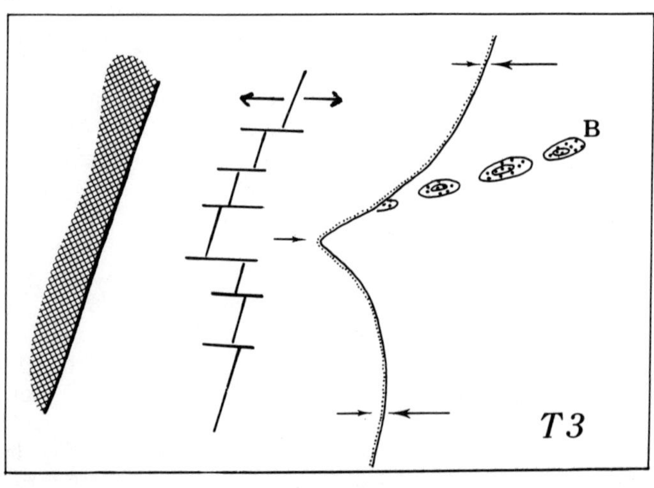
B
T3

tions for ridges whose forward edge has just arrived at the trench.) Cusps are, therefore, not generally expected where an oceanic plate underthrusts a continental one. A cusp, once formed, could survive long after the ridge has been subducted. Therefore, some ridgeless cusps are to be expected. Since cusp formation requires relative rotation of arc materials, paleomagnetic tests are possible. A ridge should roughly parallel the relative motion between the consumed and consuming plates, otherwise the intersection may not persist long enough at one site to produce a cusp. "Long enough" depends on rates of back-arc extension and subduction. Besides the mechanism of Figure 3, the plates could respond to the aseismic ridge by (a) reversing the polarity of the arc (Karig and Mammerickx, 1972), (b) breaking apart of the downgoing plate, and (c) breaking apart of the overriding plate. Mechanism (b) has been suggested by van Andel and others (1971) for the Cocos Ridge and mentioned by Karig and others (1973) for the Benham Rise. Mechanisms (a) and (b) are also considered in several cases discussed in this paper. Some explanation is required for (a): oceanic ridges are generally intermediate in crustal thickness between continents and normal ocean crust (Vogt, 1973a). The ridges are usually not large and buoyant enough to cause arc polarity reversal (Karig and Mammerickx, 1972) (Fig. 2) and thereby save themselves from being recycled into the mantle. However, we postulate that they have enough buoyancy to modify the trace of the subduction zone (Fig. 3). It seems conceivable, however, that very large oceanic platforms with thick crusts may behave as microcontinents and resist subduction by inducing arc polarity reversals when they enter the subduction zone (Kroenke, 1974). Some protocontinents might then also have been nucleated by aseismic oceanic platforms or ridges, not only by island arcs as heretofore suggested. Indeed, the crustal

---

Figure 3. Schematic time sequence of a "semi-unsubductable" aseismic ridge on an oceanic plate approaching a subduction zone, behind which back-arc extension is forming a marginal basin. As massive part of buoyant aseismic ridge enters trench, it resists subduction and thereby also locally retards eastward advance of arc, thus creating a cusp or indentation in the plate boundary. Note that ridge does not need to parallel subduction vector, and therefore arc-ridge intersection can migrate along the strike of the arc. If ridge is a single large mass, it may create a cusp and then be subducted, leaving an "empty" cusp for the observer. Back-arc extension is shown as a well-defined spreading axis, although this is neither proven to be generally true, nor necessary for cusp formation. If extension is caused by simple spreading, transform faults would probably develop as shown, to accommodate variations in spreading rate. Continuous subduction of a massive aseismic ridge not parallel to the subduction vector would be expected to produce a complex series of cusps as the ridge-trench intersection migrated along the trench (north in this figure). Modified from Vogt (1973a).

structure of many aseismic ridges resembles that of the deeper continental crust (Kroenke, 1974).

In most cases we will show the geological data support our basic premise that the subduction zone was affected by a *pre-existing* aseismic ridge. There are cases, such as the Tehuantepec Ridge, where the feature has been postulated to be an *effect* of the subduction process (Truchan and Larson, 1973). We cannot always exclude this type of mechanism, but consider it generally inadequate.

Caroline and Marcus-Necker Ridges and Marianas-Bonin Arcs

The argument for extension behind the Marianas and Izu-Bonin arcs (Figs. 4, 5) is probably more convincing than for any other island-arc–marginal-basin system (Karig, 1971a, 1971b; Moberly, 1972). It is also here that the proposed effect of aseismic ridges on the evolution of such a system can readily be visualized, if not finally proven (Fig. 6). According to Karig, the most recent pulse of back-arc extension opened the Mariana Trough and Bonin zone (trough) in latest Tertiary and Quaternary time. High heat flow, thin sediments, and elevated ocean floor all attest to the youth of these basins. The Parece Vela Basin to the west is believed to have opened in early Miocene time (Karig, 1971a); Karig's bathymetry suggests, as earlier data did to Hess (1948), that the Yap Trench has an inactive continuation, in the form of irregular ridges and deeps, through the Parece Vela Basin. This basin can, therefore, itself be divided into two parts. We shall use this observation later. Karig interprets the Palau-Kyushu Ridge as a remnant island arc, split from the West Mariana Ridge as a result of the postulated early Miocene pulse of extension. Basement depths and heat flow in the Parece Vela Basin are consistent with the proposed timing, but spreading-type magnetic lineations have not been identified either here or in the younger Mariana Trough. Perhaps the process of extension is broader and more complex, without a sharp central axis of lithosphere accretion (Karig, 1971a).

The hypothesis illustrated in Figure 6 implies that the Marianas Ridge acquired its present "bulge" in post-Miocene time. Therefore the rocks at the northern and southern ends of the arc must have been deformed counterclockwise and clockwise, respectively. Paleomagnetic evidence from southern Guam indicates 50° to 60° clockwise roation in post-Miocene time, thus supporting our hypothesis (Larson and others, 1975).

West of the Palau-Kyushu Ridge lies the still older and deeper West Philippine Basin. We shall briefly reconsider the ridges of this basin later, with reference to the shape of the *western* boundary of the main Philippine plate.

We now turn to the aseismic ridges on the main Pacific plate and their relation to the complex history of eastward migration of the Marianas and Izu-Bonin frontal arcs. An irregular assortment of aseismic ridges and seamounts presently lies seaward of these arcs (Figs. 4, 5). The two most prominent cusps in the line of arcs are the northern and southern ends of the Mariana arc, and the two most

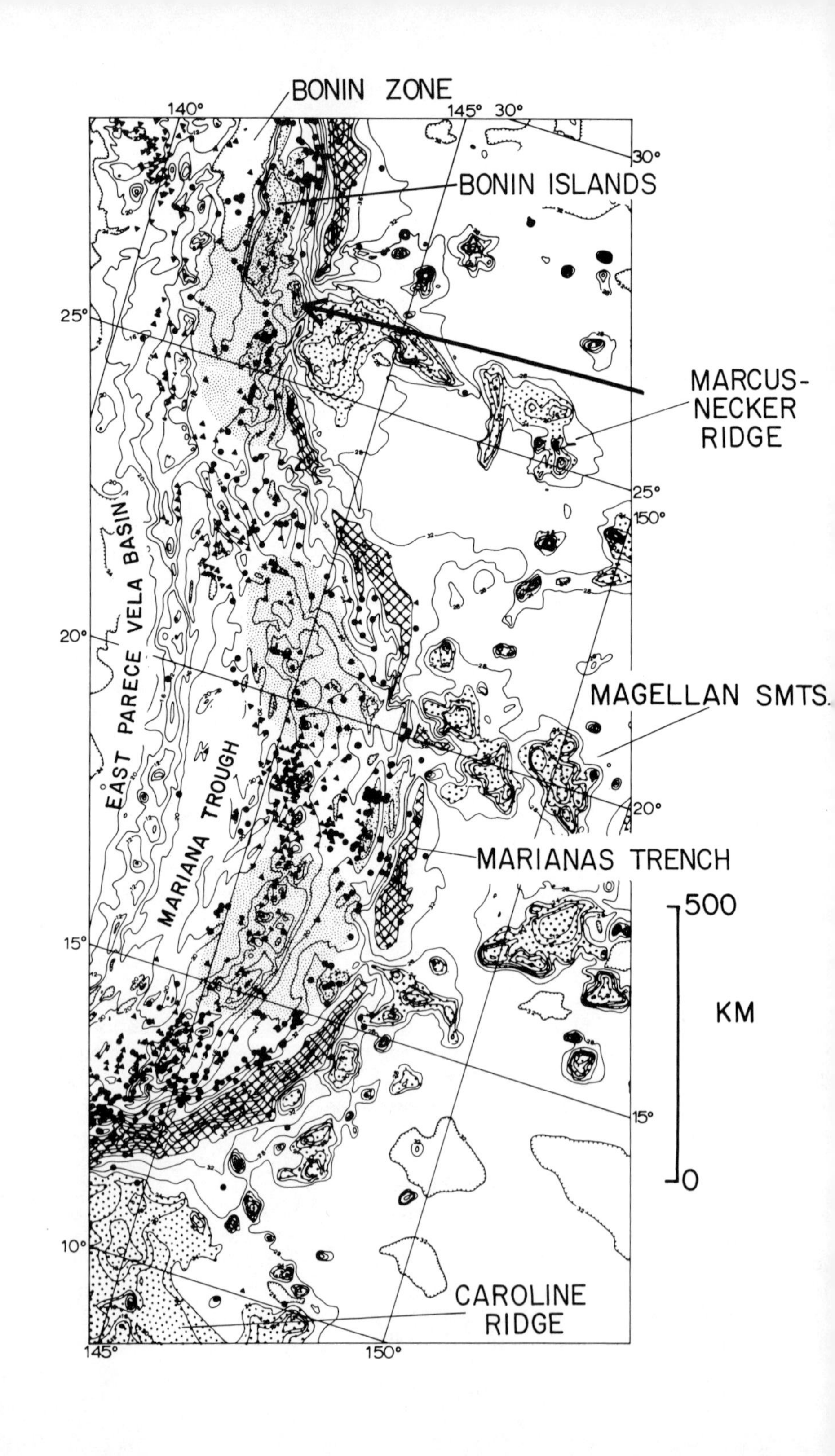

BONIN ZONE
140°
145°
30°
30°
BONIN ISLANDS
25°
MARCUS-
NECKER
RIDGE
25°
150°
EAST PARECE VELA BASIN
20°
MARIANA TROUGH
MAGELLAN SMTS.
20°
MARIANAS TRENCH
500
KM
0
15°
15°
10°
CAROLINE
RIDGE
145°
150°

massive aseismic ridges on the adjacent Pacific plate, the Marcus-Necker and Caroline Ridges, presently trend into these cusps (Vogt, 1973a). The Marcus-Necker Ridge, or Mid-Pacific Mountains (Fig. 4), is a prominent but poorly known seamount chain reminiscent of the Hawaii-Emperor Ridge except probably older, about mid-Cretaceous (Menard, 1966; Morgan, 1973). More irregular groups of generally smaller seamounts, also of probable Mesozoic age (for example, the Magellan Seamounts) lie in front of the central Marianas arc. These smaller seamount groups disturb the details of Mariana Trench topography but have apparently not interfered with the eastward motion of the arc (Fig. 4).

The Caroline Ridge is an enigmatic feature, possibly of multiple origin (Fig. 5). Its western end, which in our view has "pinned" the southern end of the Marianas arc, is a broad arch incised along its crest by the Sorol Trough, a graben-like feature possibly the site of recent ocean-floor spreading. Broad areas of the Western Caroline Ridge are less than 2.5 km deep. Farther west the Caroline Ridge becomes a chain of seamounts and islands. The Eastern and Western Caroline Ridges may be genetically unrelated (Bracey and Andrews, 1974), although Clague and Jarrard (1973) ascribe the entire Caroline Ridge to post-middle Tertiary motion of the main Pacific plate over a fixed hot spot. Deep drilling on the ridge and Caroline basins suggests that they were formed in Oligocene time (Winterer and others, 1971; Fisher and others, 1971). The Caroline Ridge problem is complicated by its position on the boundary between the Jurassic Pacific plate and the much younger crust of the Caroline basins, which apparently formed by spreading from an east-west axis in Oligocene-Miocene times (Moberly, 1972; Vogt and Bracey, 1973). The Caroline Ridge area is tectonically active, as evidenced by widely spaced epicenters (Katsumata and Sykes, 1969). We believe that the Sorol Trough and similar fractures to the southeast may reflect incipient breakup of the southwestern corner of the Pacific plate, perhaps because of the mechanical difficulties involved in subducting the massive Caroline, Eauripik, and Ontong-Java Ridges. Similar widespread seismicity in the eastern Cocos plate may also reflect stresses created in the plate by attempted subduction of the

---

Figure 4. Bathymetry (Anonymous, 1970b) and seismicity (Anonymous, 1972b) of the Marianas arc area. Contours at 400 fm (740 m) interval. Aseismic ridges and seamounts shown by wide-spaced stippling; relatively aseismic areas (1961–1971) by fine stippling; Marianas and Bonin Trenches by cross-hatching; and exceptional highs on mid-slope ridge or terrace, by irregular stippling. The last mentioned may be due in part to scraped-off seamounts. Earthquake symbols are: 0–70 km focal depth, closed circles; 71–300 km, upward-pointing triangles; greater than 300 km, downward-pointing triangles. Vector shows last 10 m.y. movement of a point on the underthrusting plate, now at the head of the arrow, with reference to the underthrust plate, as computed from relative motion poles of Morgan (1973). Vector neglects additional motion of frontal arc due to back-arc or inter-arc extension (Karig, 1972).

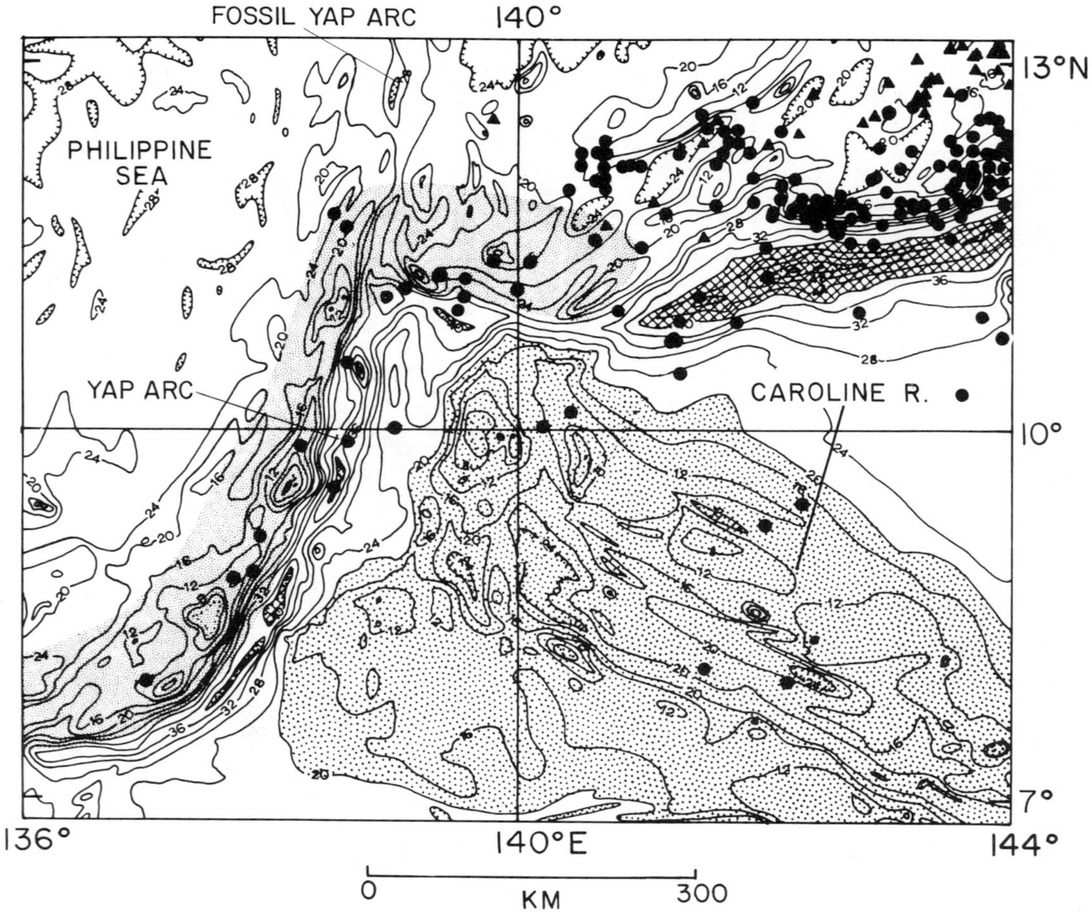

Figure 5. Bathymetry and seismicity of intersection of Caroline Ridge with Yap and Marianas arcs. Symbols as in Figure 4.

Figure 6. Schematic hypothesis of Neogene evolution of Marianas arc area, using concept of Figure 3 and data of Figures 4 and 5. Based in part on Moberly (1972), Karig (1971a, 1971b, 1972), Vogt and Bracey (1973), Fisher and others (1971), Winterer and others (1971), Bracey and Andrews (1974), and Uyeda and Miyashiro (1974). Active subduction zones are solid teeth; remnant arcs (Karig, 1972) are open teeth. Caroline Ridge is shown formed more or less simultaneously by a hot spot which broke the old Pacific plate and subsequently formed the Eauripik Rise. The origin of the Caroline Ridge is not of vital importance to the thesis of this paper, however.

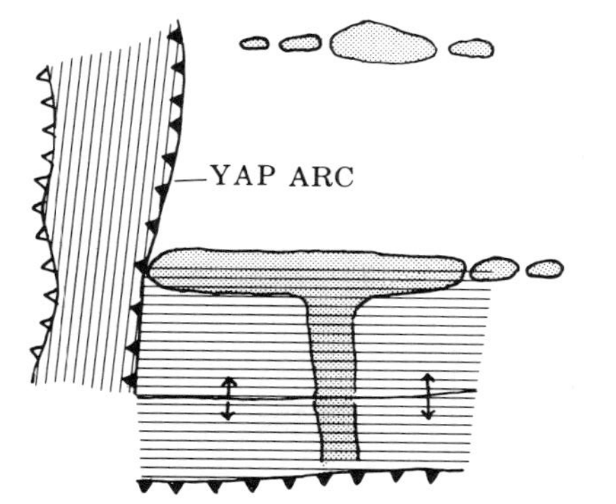

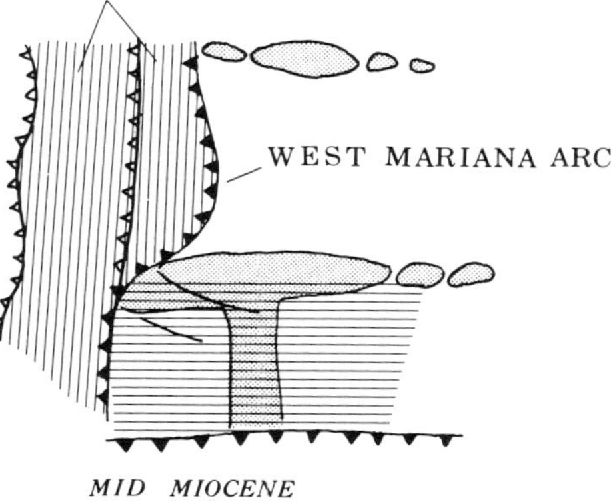

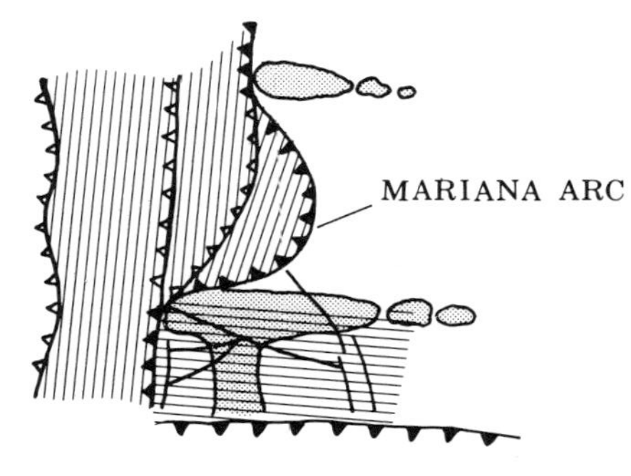

Cocos Ridge (van Andel and others, 1971). Alternative hypotheses attribute the Sorol Trough and similar fractures to tensional rifting of the descending lithosphere (Andrews, 1971) or back-arc spreading related to a former north-dipping subduction zone (Bracey and Andrews, 1974). The main point is that the Caroline Ridge is Oligocene or early Miocene in age (whether or not older Pacific crust lies beneath it). Since the Parece Vela Basin opened in early to mid-Miocene time (Karig, 1971a), we propose that the trench made contact with the ridge sometime in this interval. The east Parece Vela Basin then developed, leaving the Yap arc as a fossil remnant. The latest episode of back-arc spreading was influenced by the Marcus-Necker Ridge as well as the Caroline Ridge. Subduction and back-arc extension along the Yap arc was further inhibited by the thin, low-density lithosphere of the Caroline Basin, which does not sink as readily into the asthenosphere as the old main Pacific plate to the north. The above complexities may have fractured the relatively young Caroline Basin crust in latest Tertiary times (Fig. 5).

We now briefly list other possible effects of subducted aseismic ridges in the Marianas area. First, a direct corollary of our previous discussion is that there should be no active marginal basins where the Caroline and Marcus-Necker Ridges intersect the subduction zone. This indeed seems to be the case (Fig. 1 of Karig, 1971b). Second, we may ask whether earthquake seismicity has been affected. There is a sharp reduction in recent seismicity between the Marianas and Yap arcs (Fig. 5), but this may reflect the much slower rate of motion between the Caroline Ridge and the main Philippine plate, as implied by Karig's (1971a) model. Earlier plate tectonic analyses (Katsumata and Sykes, 1969) would predict reduced convergence, hence reduced seismicity, but they do not account for the *abruptness* of the change. Upon subduction, the younger, thinner, and hotter Caroline Basin lithosphere may also fracture less than the old Pacific plate when being subducted.

Farther north, the intersections of the Marcus-Necker Ridge and two lesser seamount groups to the south with the Marianas arc seem to be accompanied by areas of reduced seismicity (Fig. 4). The effects are only slight, however, and additional time is needed to verify that these are permanent features. If they are, it would mean the lithosphere containing seamount chains is more ductile or perhaps also thinner, and more likely to deform by creep than by brittle fracture.

Are seamounts ever sheared off the downgoing plate? There are some hints to this effect in the Marianas area. The eastern flank of the frontal Marianas-Bonin arc slopes steeply near sea level but flattens eastward to form a bench or subsidiary ridge (Karig, 1971a). This mid-slope basement high is a ridge or series of elongate highs; the Bonin Islands represent the unusual case of this ridge breaching sea level. Clearly, the mid-slope ridge belongs to the overriding plate; probably the upper part consists of scraped-off pelagic sediments as well as arc-derived volcanogenic materials. A number of peaks on the mid-slope ridge have the dimensions of seamounts and occur landward of observed seamount groups (Fig. 4). The unusual Bonin Islands Ridge in particular seems to be associated with the intersection of the Marcus-Necker Ridge with the trench during the last 10 m.y. or so, if the relative motion in this area is approximately correct (Fig. 4). Perhaps sheared-off seamounts or portions of seamounts have been emplaced at depth below these mid-slope peaks. An immense free-air gravity high of +350 mgal over the Bonin Ridge (Hayes and Ewing, 1970) certainly indicates a large, isostatically unsupported mass excess. Stratigraphic dating implies that the Bonin Ridge was formed in early Tertiary time since intermittent shallow water deposits of this age occur there (Karig, 1971a). Detailed geophysical work along the mid-slope ridge is required to demonstrate that sheared-off seamount volcanics or their mantle roots occur at depth below the Bonin Islands.

Other Aseismic Ridges in the Western Pacific

We now briefly consider miscellaneous other ridges on the Pacific, Philippine, and Indian plates in relation to the shape of the subduction zones. Areas to be discussed (Fig. 1) are as follows: Two cusps east of Japan, the western edge of the Philippine plate, the complex part of the Pacific plate boundary from New Guinea to Fiji, and the Tonga-Kermadec "arcs" (Fig. 7). The possible significance of the Chagos-Maldive and Ninetyeast Ridges was discusssed by Vogt (1973a) and will not be elaborated upon here.

Two conspicuous but relatively subdued cusps east of Japan are probably the best examples of cusps *without* prominent aseismic ridges trending into them (Fig. 1). There are irregular groups of widely scattered, relatively small seamounts that one could associate with these cusps, but it is doubtful that such minor features could have caused the cusps. At best, the small peaks could be the trailing ends of a former, much more massive ridge subducted in the past. Of course, our model (Fig. 3) does not require all cusps to have ridges, but only the converse of this. (If the ridge just reached the trench, of course, no cusp is expected either.) In any case, the southern of the two cusps is a triple junction for the Pacific, Philippine, and Eurasia plates and probably should not be "explained" by the simple hypothesis of Figure 3. Both cusps may be as young as Miocene, since heat flow and other data suggest the Parece Vela, Japan, and Okhotsk Basins behind these cusps are that young (Karig, 1971b; Uyeda and Miyashiro, 1974; Moberly, 1972). There are prominent variations in seismic activity along the subduction zone, but it is not clear that any of the relatively "aseismic" areas are associated with subducted seamount chains.

The western margin of the Philippine plate is indented by a number of cusps and other irregularities (Fig. 1); all appear to be associated with aseismic ridges and thus support our model. The northernmost case is the indentation into the northwestern Nankai (Southwest Japan) Trench by the northern extension of the Bonin Ridge, actually a pair of ridges called Shichito-Iwojima and Nishi Shichito (Fig. 10 of Uyeda and Miyashiro, 1974). The plate boundary east of this minor cusp is the Sagami Trough, apparently mainly a transform fault (Fitch, 1972). To the southwest, the Palau-Kyushu Ridge and the Amami Plateau extend into the cusp between the Ryukyu and Nankai (Southwest Japan) Trenches (Fig. 1). The Daito and Oki-Daito Ridges appear to be just approaching the Ryukyu Trench and have not produced a dent in it (Fig. 10 of Uyeda and Miyashiro, 1974). The

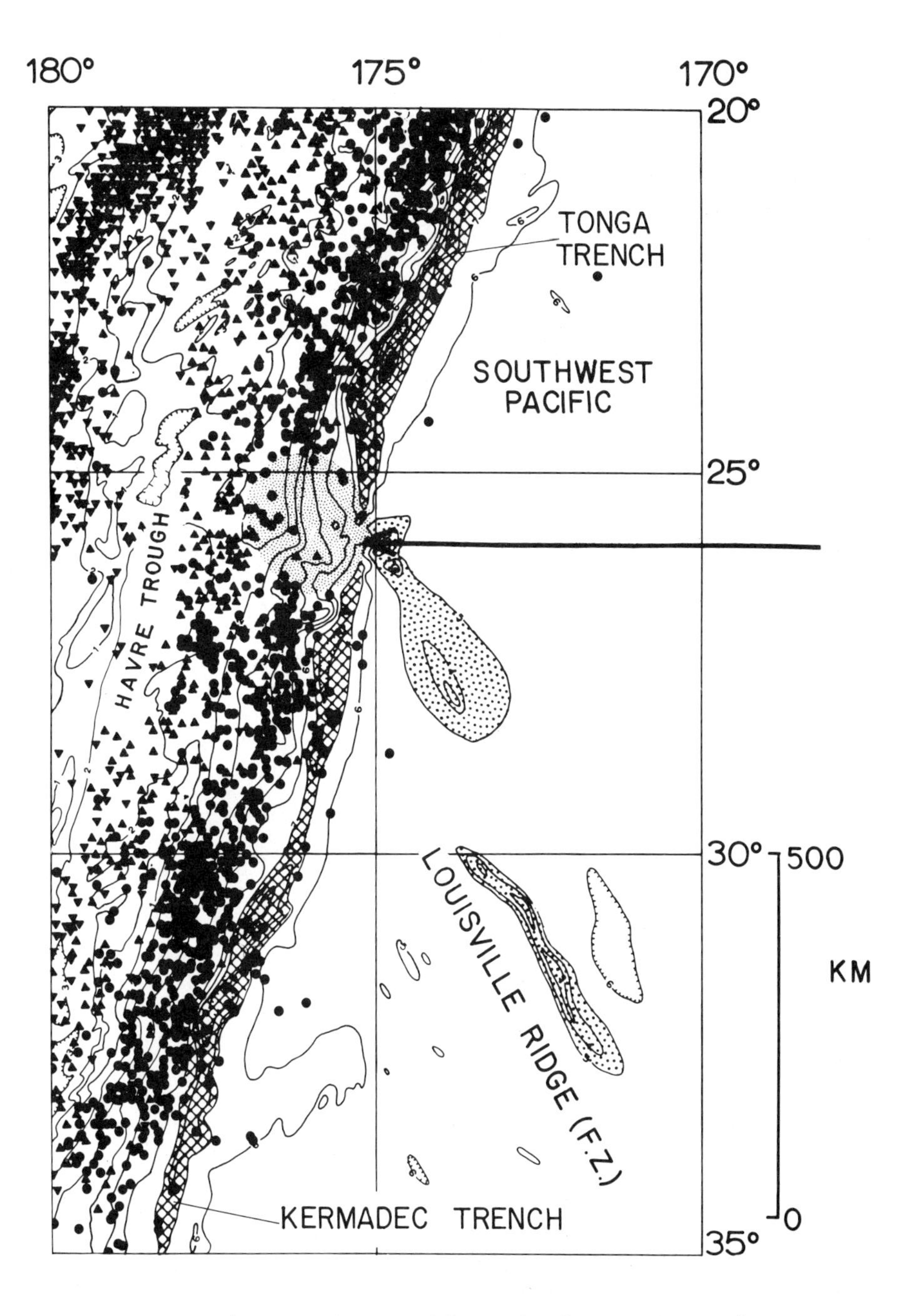

Figure 7. Bathymetry (Mammerickx and others, 1971) and seismicity, intersection of Louisville Ridge with Tonga and Kermadec "arcs." Symbols as in Figure 4.

Palau-Kyushu Ridge may be a remnant island arc (Karig, 1971a, 1971b), and perhaps also the Daito and Oki-Daito Ridges (Karig, 1972).

Between the Ryukyu Trench and the Philippine Trench there is a complex westward bulge in the main Philippine plate. This bulge actually consists of two cusplike indentations in the plate margin (Fig. 1). The Central Basin fault or Philippine Ridge may extend as far as the northern cusp, located east of Taiwan, and the southern cusp is occupied by the Benham Rise. The latter feature is called the Daito-Luzon Rise by Uyeda and Miyashiro (1974) and the Anson Massif by Soviet authors (Anonymous, 1970a). There is possibly a third aseismic ridge involved in the bulge—the Palaui Ridge (Karig, 1973) which roughly parallels the plate boundary and lies on the extreme western edge of the main Philippine plate, between the two cusps. According to Karig, this ridge may be an early Tertiary mid-slope basement high like the Bonin Island Ridge (Fig. 4), or else an old remnant arc.

The Central Basin fault, or Philippine Ridge, may be an extinct segment of the Mid-Oceanic Ridge (Ben-Avraham and others, 1972; Uyeda and Ben-Avraham, 1972; Uyeda and Miyashiro, 1974). Recent deep drilling led Karig and others (1973) to revive the older idea that the Philippine Ridge is, after all, a major transcurrent fault. However, Louden and Sclater (1974) claimed to have found a symmetrical magnetic anomaly pattern that would imply that the ridge is a former spreading axis that ceased its activity about 50 m.y. B.P. The feature appears to be a series of northwest-trending ridges and valleys; locally a riftlike valley occurs at the axis. Although its origin is not critical to our hypothesis, it is worth noting that this might be the only example of an extinct spreading axis extending into a cusp. Uyeda and Miyashiro (1974) showed this feature terminating well east of the cusp at the southern end of the Ryukyu Trench; however, a Soviet tectonic chart (Anonymous, 1970a) showed it extending precisely into the cusp. Karig (1973) also indicated that the ridge extends all the way, its northwestern end having been buried by sediment and therefore no longer apparent in the bathymetry.

The Benham Rise (Karig, 1973) is a broad swell consisting of north-south ridges covered with pelagic sediments exceeding 500 m in thickness. The rise reaches 1 to 3 km above the normal depths of the West Philippine Basin. This thickness of pelagic sediments shows that it is not a *result* of recent subduction (Karig, 1973), and, whatever its origin, it has been available to influence the geometry of such subduction.

Because of the complexity of the region and the paucity of data (Karig, 1973), it would be premature to guess in detail how the plate boundary adapted itself to these two aseismic ridges. However, several observations are relevant to the question of whether the ridges *could* have exerted such an influence. First, the Philippine Ridge (Central Basin fault) roughly parallels the relative motion vector between the main Asia and Philippine plates, at least since late Tertiary time (Fitch, 1972; Karig, 1973). The ridge has, therefore, trended into the southern end of the Ryukyu Trench for some time, a requirement of our hypothesis. Second, there is a back-arc basin, the Okinawa Trough, which terminates near Taiwan. This trough is young and may be still widening and, therefore, allows the Ryukyu arc to migrate seaward with respect to the Philippine Ridge-cusp junction. The Philippine Ridge is not a major topographic high, however, and it is not easy to visualize a cusp being formed by buoyancy alone (Fig. 3).

The evolution of the plate boundary from Taiwan to the Philippines is much more complex, and it is difficult to evaluate the possible role of the Benham Rise or its possible now-subducted extensions, and the Palaui Ridge. Karig (1973) believed that a west-dipping Benioff zone existed from east of southern Taiwan down to east of Luzon. This was the situation in Late Cretaceous and early Tertiary time; opening of the South China Basin is supposed to have occurred in conjunction with this subduction (Karig, 1973), although the high heat flow in the basin would suggest that it is younger (Karig, 1972). By late Miocene time, an arc polarity reversal is supposed to have occurred; south of Taiwan, underthrusting from the west then began to destroy the South China Basin and, even more recently, began to cause the arc to collide with Taiwan. Perhaps the entry of the Philippine, Benham, and Palaui Ridges into the former west-facing subduction zone in the Miocene helped cause the arc polarity reversal. Subduction from the east may be beginning anew, at least east of Luzon; the Benham Rise would have influenced the location and modification of this subduction zone, which may have been active at a subdued rate for a longer period of time than proposed by Karig. Puerto Rico presents a somewhat similar setting as Luzon, located as it is between the active Puerto Rico and Muertos Trenches since middle or late Tertiary time (Matthews and Holcombe, 1974). If an east Luzon subduction zone has existed for some time, it is not hard to envisage the role of the Benham Rise in creating a notch in the plate boundary. Any amount of back-arc extension in this Luzon microplate, between the trenches on either side, would provide the degree of freedom to develop such a notch. One can also consider more complex interactions involving strike-slip faulting, as in the Hispaniola area (to be discussed later).

Another possibility is that the relative unsubductibility of the Benham Rise caused the Philippine Ridge (fault) to form (Karig and others, 1973). This would be analogous to formation of the Panama fracture zone by the jamming action of the Cocos Ridge (van Andel and others, 1971). We do not favor such an explanation for the Philippine Ridge because it implies seemingly excessive strengths for the lithosphere. Besides, it would require that the Benham Rise already arrived at its present site next to Luzon some time between the Eocene and Miocene, since that appears to be the age of the Philippine Ridge (fault) (Karig and others, 1973).

We now turn to another complex plate boundary—the region between New Guinea and the northern end of the Tonga Trench. A regional kinematic synthesis appears in Moberly (1972). Among other recent papers dealing with plate motions along this complex boundary between the Pacific and Indo-Australian plates are those of Luyendyk and others (1974), Chase (1971), and Karig and Mammerickx (1972) for the New Hebrides–Fiji area; Karig (1972), Milsom (1970), Luyendyk and others (1973), Krause (1973), Vogt and Bracey (1973), and Bracey (1975) for the New Guinea–Solomons area; and Johnson and Molnar (1972) and Fitch (1972) for the entire region.

The Eauripik Rise is a broad, slightly arcuate feature that trends approximately north-south from the Caroline Ridge to an east-west trench system north of New Guinea. The crest of the rise is 2 to 3 km in depth and thus represents about 2 km relief above the adjacent West and East Caroline Basins. Seismic reflection profiling and deep drilling (Winterer and others, 1971) indicate that the Eauripik Rise was formed at the same time (Oligocene) as the Caroline Basins. The rise is supported by thickening of layer three from 3 or 4 km under the basins to perhaps 10 to 15 km under the crest (Den and others, 1971). The origin of the rise is still obscure; a hot-spot–generated aseismic ridge like the Iceland-Faeroe Ridge is one possibility (Fig. 6). The Eauripik Rise trends into a gentle cusp between the New Guinea Trench south of the West Caroline Basin and the Manus Trough or West Melanesian Trench. The New Guinea Trench is the site of present underthrusting of the Pacific plate (or Caroline subplate) under the Indo-Australian plate. To the east, there are believed to be three small plates between the two major ones—the North and South Bismarck plates and the Solomon Sea plate (Johnson and Molnar, 1972; Krause, 1973; Karig, 1972; Luyendyk and others, 1973). The relative motions among these minor plates is not well known, and there is considerable disagreement about the tectonic role played by the various features in this area (Karig, 1972; Luyendyk and others, 1973). Of interest here is that the two Bismarck plates are sandwiched between a more active north-dipping New Britain zone and a south-dipping West Melanesian subduction zone (Karig, 1972). It thus appears that the Bismarck Basin could be the result of back-arc spreading, allowing the West Melanesian arc to migrate north. However, left-lateral strike-slip solutions (Johnson and Molnar, 1972) suggest that the West Melanesian arc is presently an oblique subduction zone, perhaps like the western Aleutian arc. We propose that the northward migration of the arc has been hindered on the west by the Eauripik Rise and on the east by the massive Ontong Java Plateau.

The Ontong Java Plateau (Moberly, 1972; Kroenke, 1974) may be an example of an aseismic ridge sufficiently "continental" to induce an arc polarity reversal. The plateau rises 2 to 3 km above the surrounding Pacific Basin, and its Cretaceous basement is covered by about 1 km calcareous sediments (Maynard and others, 1974). Its crust is greatly thickened compared to the oceanic average (Kroenke, 1974). The Ontong Java Plateau presently fits into a gentle southward bend of the Pacific plate—the Solomon arc. In this case, the aseismic ridge is on the underthrust plate (Johnson and Molnar, 1972) so that the simple model of Figure 3 does not apply. However, an inactive trench and a remnant arc lie north of the present Solomon arc (Karig, 1972) so that the possibility of geologically recent southward underthrusting, followed by an arc polarity reversal due to the Ontong Java Plateau, appears inviting.

Farther to the southeast, the Fiji Plateau seems to occupy a major southward salient of the Pacific plate (Fig. 1). In fact, however, there is now ample evidence that the plateau is not an aseismic ridge, but a region of late Tertiary to Holocene ocean crust containing one or more small subplates. The plateau is 2 to 3 km deep, exhibits high heat flow (Sclater and Menard, 1967), very thin sediment, and suggestions of short north-trending spreading axes and other plate boundaries (Chase, 1971). At present, the Australian plate is underthrusting the main Pacific and Fiji subplates along the New Hebrides Trench, but it has been suggested that the New Hebrides arc once extended east-west and faced south, but then reversed its polarity in Miocene time (Karig and Mammerickx, 1972; Chase, 1971). Alternative scenarios are postulated by Luyendyk and others (1974). If the arc reversed polarity, perhaps aseismic ridges on the main Pacific plate were contributory factors. For example, if the Pacific plate has been moving past the Australian plate at rates of about 10 cm/yr (Chase, 1971), the massive Ontong Java Plateau would have been opposite the New Hebrides arc at the time (Miocene) of arc polarity reversal, as postulated by Karig and Mammerickx (1972).

A later oblique but also south-dipping subduction zone along the Vitiaz Trench could have been immobilized by another aseismic ridge, the Border Plateau. This elongated rise, with crestal depths of less than 2 km, presently abuts the fossil Vitiaz Trench from the east and would have made contact with this trench when the latter became inactive, about 10 m.y. B.P. according to the scenario depicted in Figure 11 of Chase (1971). It is not yet certain the aseismic ridges on the main Pacific plate played a role in the complex development of the Fiji area; however, there must be some reason for the plate boundaries to evolve as they did, and we offer the aseismic ridges as one possible mechanism.

Following the boundary of the Pacific plate southeastward, we come to the remarkably linear Tonga-Kermadec "arcs." As in the case of the Marianas arc, there has been late Tertiary to Holocene back-arc extension behind the Tonga and Kermadec arcs (Karig, 1970a, 1970b; Sclater and others, 1972). The Lau Basin represents about 250 km of inter-arc extension behind the Tonga Trench; the basin widens to the north. Behind the Kermadec Trench lies the somewhat narrower (100 km) Havre Basin, which continues south into New Zealand (Karig, 1970a). The general northward increase in back-arc extension from New Zealand to the northern end of the Tonga Trench may reflect (a) the northward increasing age and thickness of the downgoing lithosphere (hence more rapid sinking of the subducted slab), or (b) earlier beginning of extension in the north, or (c) the more rapid subduction in the north (Sykes and others, 1970).

The uncomplicated characteristics of the Tonga-Kermadec "arcs," unlike the subduction zones from Fiji to New Guinea, were to be expected by our hypothesis (Fig. 3), since no *major* aseismic ridges exist on the downgoing Pacific plate east of the Tonga-Kermadec trenches. However, there is the relatively minor Louisville Ridge, evidently associated with a major transform fracture zone (Hayes and Ewing, 1971) that intersects the junction between the Tonga and Kermadec trenches (Fig. 7). The Louisville Ridge is 25 to 50 km wide and rises 1 to 3 km above the surrounding later Cretaceous basin. Additional data may show it to be more linear than indicated (Fig. 7). The trenches are offset slightly and form a sill where the Louisville Ridge intersects the arc; a small but prominent region of reduced seismicity also exists at the intersection. Sykes and others (1970) first mentioned the possibility that a feature like the Louisville Ridge on the downgoing plate might be involved in seismic discontinuities in the intersection area. In Figure 8, we have projected the extension of the Louisville Ridge onto Sykes and others' vertical projection of seismic activity (Fig. 19 of Sykes and others, 1970). In making the projection, we observed the contortions of the plate surface implied by the seismicity: this is why the projected Louisville Ridge in

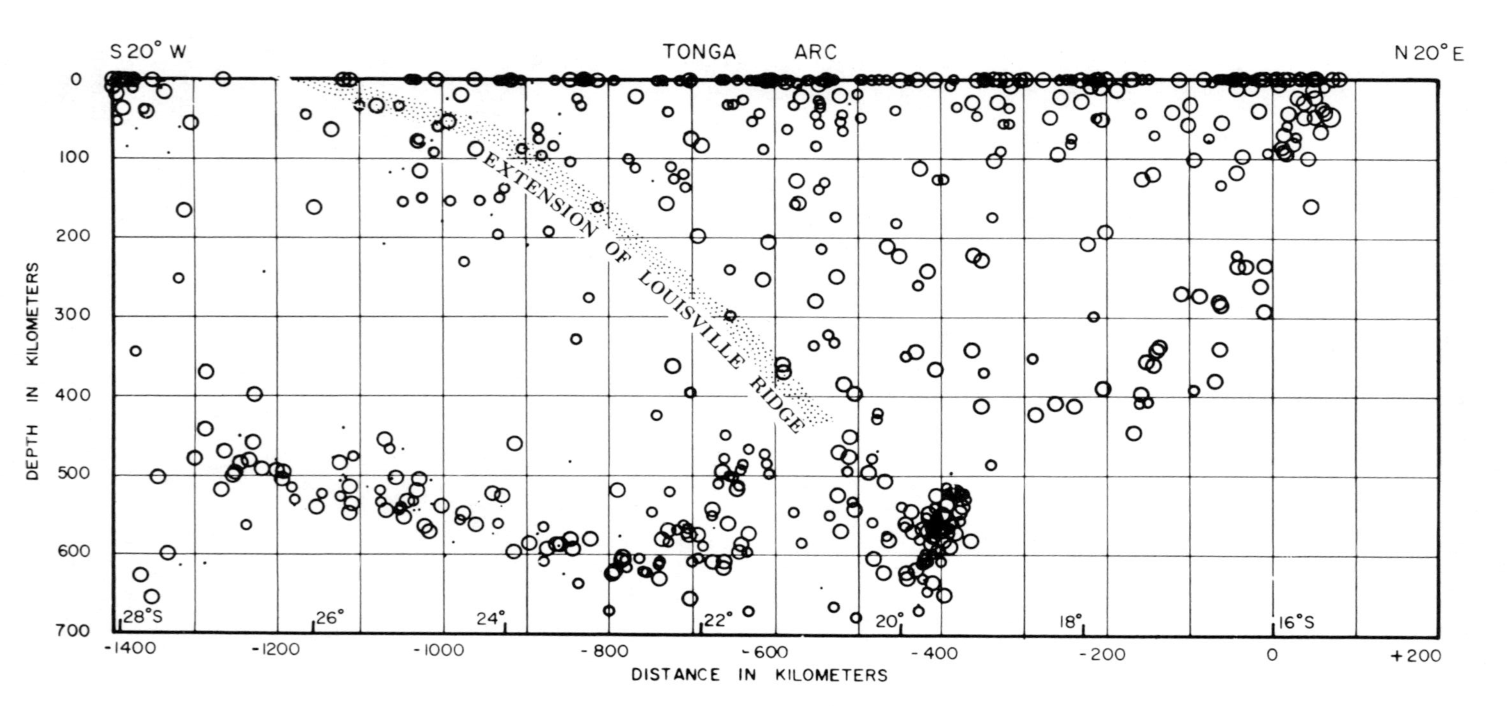

Figure 8. Projection of Louisville Ridge (stippled) onto downthrust Benioff zone, itself projected on a vertical section by Sykes (1966) and Sykes and others (1970). Ridge projection is curved because slab is contorted. Larger symbols denote more accurately determined hypocenters.

Figure 8 is not a straight line. The projection clearly shows the aseismic region where the ridge intersects the surface. The ridge also seems to bound a region of reduced intermediate depth seismicity (on the left) and finally approaches the major slab contortions between km 400 and 700. It may be that the fracture ridge is a line of weakness that has localized warping of the downgoing plate, a warping in response to the curvature of the Earth (Frank, 1968; Scholz and Page, 1970).

Emperor and Gulf of Alaska Seamount Chains and Aleutian Arc

The Aleutian arc is one of the major arclike subduction zones in the world ocean, a zone along which the Pacific plate underthrusts the Americas plate (Fig. 1). The arc is bounded on the west by a cuspate junction with the Kuril-Kamchatka arc; its eastern junction lies in south-central Alaska, where the north-east trends of the arc make an abrupt 90° turn to the south-southeast. A major aseismic ridge, the Hawaii-Emperor Seamount chain, trends into the Kuril-Aleutian cusp (Fig. 9), and three lesser seamount chains trend into the Aleutian Trench near the eastern cusp. The main, southward convex bulge of the arc thus lies between two groups of aseismic ridges on the downgoing plate, and we therefore ask whether the geometry of the Aleutian arc could have been shaped by these ridges, along the lines postulated in Figure 3.

The Hawaii-Emperor chain is not only the most prominent intra-plate aseismic ridge on the Pacific plate, but the three-way junction between the Kuril-Kamchatka and Aleutian arcs and the Emperor Ridge is also one of the most spectacular examples of an aseismic ridge trending into a cusp (Fig. 9). The ridge is actually a chain of inactive shield volcanoes ranging in age from about 45 m.y. B.P. at the "Hawaii-Emperor Bend" (Clague and Jarrard, 1973) to slightly more than 72 m.y. B.P. (Meiji Guyot, Fig. 9) (Scholl and Creager, 1973). The chain appears to have been formed during a phase of northerly motion of the Pacific plate over the Hawaii hot spot (Morgan, 1971, 1973). Approaching the Aleutian-Kamchatka cusp from the south, the Emperor Ridge strikes N15°W to about 51°N, 168°E, where it appears to widen substantially and change trend to N45°W. This northwesterly trend nearly parallels the late Tertiary relative motion between the Americas and Pacific plates (Fig. 9); it follows that the Emperor chain may have maintained its intersection with the Aleutian-Kamchatka cusp for some time—perhaps several tens of millions of years. Such a sustained intersection seems to be necessary, by the hypothesis of Figure 3, because the Aleutian arc and the basins behind it certainly cannot have formed in the last few million years. Although the present chain of active andesitic volcanoes along the Aleutian Ridge represents an outbreak of volcanism that began only 2 m.y. B.P., the presence of older sedimentary and igneous formations indicates that the ridge probably formed in early Tertiary time (Scholl and others, 1970; Burk, 1965). The few fossil dates there are do not preclude an age as young as Oligocene for the Aleutian Ridge west of Adak (Grow and Atwater, 1970). There

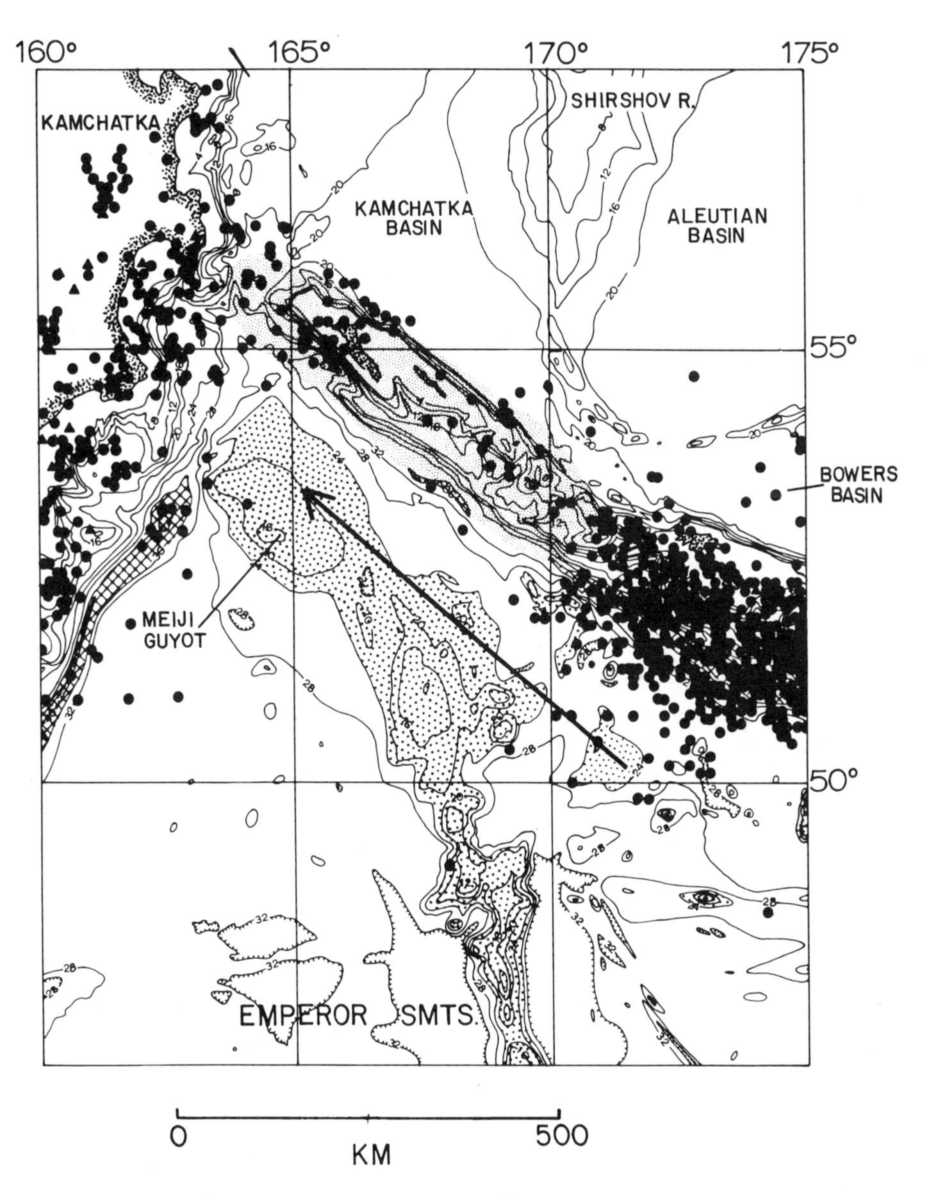

Figure 9. Bathymetry (Anonymous, 1971b) and seismicity, intersection of Emperor Ridge (seamounts) with Aleutian-Kamchatka cusp. Symbols as in Figure 4.

is evidence for a Cretaceous subduction zone whose rocks are now exposed on the islands south of the Alaska Peninsula (Burk, 1972); this fossil arc lies along the edge of the Bering shelf (Moore, 1973); post-Cretaceous inter-arc spreading and southwestward arc migration could then have opened the main Aleutian Basin. Accelerated deformation, uplift, and plutonism in the middle to late Miocene may reflect more rapid underthrusting of the Aleutian arc by the Pacific plate at that time; this Miocene event followed a quiescent interval of slow plate motion from about 45 to 25 m.y. B.P. (Scholl and Creager, 1973; Clague and Jarrard, 1973).

Although at least portions of the Aleutian Ridge already existed in the lower Tertiary, this does not prove that the present shape of the arc was the same then. Other arcs are considered to have acquired their curvature by inter-arc or back-arc spreading (Karig, 1970a, 1970b, 1971a, 1971b). The Marianas, Bonin, Tonga-Kermadec, and Scotia arcs have probably all been distorted to their present configuration *subsequent to the oldest formations in the arc ridge.* If such inter-arc extension occurred behind the Aleutian arc, it follows that the present "arc" shape may be substantially younger, in part at least, than the lower Tertiary. A younger bending of the Aleutian Ridge might be demonstrated paleomagnetically; there should also then be post-lower Tertiary inter-arc–type extension of the Aleutian, Bowers, and Kamchatka Basins that lie north of the arc. Evidence for such extension is ambivalent at present. The relatively deep Aleutian Basin contains 3 to 4 km of sediments (Kienle, 1971) and exhibits no obvious clues to recent extension, such as high heat flow or faulting of surficial layers. In fact, Scholl and others (1974) have suggested that some north-trending magnetic lineations in the basin represent the Lower Cretaceous Keathley sequence of magnetic reversals. They suggest that the Aleutian arc formed abruptly in the middle of an existing oceanic plate, rather than by a Karig-type process (Fig. 3). If so, the arc itself may be much younger than the Aleutian Basin, and its formation may still have been controlled by the presence of the massive Emperor Ridge in the downgoing plate. However, the age of the Aleutian Basin is far from proven; deep drilling has not penetrated below the Neogene (Scholl and Creager, 1973). It will be difficult to prove that no late Tertiary extension occurred in the Aleutian Basin.

The Kamchatka and perhaps also the Bowers Basins appear to be younger than the Aleutian Basin. This is evidenced by higher heat flow, thinner sediment covers, and a drill station that recovered 30 m.y. B.P. basalt in the Kamchatka Basin (Scholl and Creager, 1973). Although back-arc extension appears to have affected those basins and may be doing so at this time, Scholl and Creager (1973) believe that the cored basalt was emplaced in a pre-existing basin, without major relative motions between the Shirshov, Bowers, and Aleutian Ridges. Kienle (1971) argued that the Shirshov and Bowers Ridges represent fossil subduction zones, with a trench located east of Bowers Ridge. He suggested that westward subduction of Aleutian Basin crust occurred during Late Cretaceous to early Tertiary times; if it continued as late as the middle Olicogene, such subduction might explain the formation of Bowers and Kamchatka Basins by back-arc spreading.

In summary, the geological and geophysical data are inadequate to determine the time when the Aleutian arc acquired its present form. At least in the western arc, the rocks are not older than early Tertiary; the arc could have developed then or subsequently. In view of the great displacement between the Pacific plate and the arc since early Tertiary time, the hypothesis that past extensions of the Emperor Ridge helped shape the arc at that time becomes less attractive, although not out of the question. The large-displacement hypothesis (Atwater, 1970) places Meiji Guyot roughly 800 to 1,000 km south of Adak Island in the middle Miocene; if the western Aleutians acquired their curvature largely in the Neogene, this would be consistent with our hypothesis insofar as a northwest-trending Emperor Ridge would already have trended into the cusp at that time. As discussed above, the geological evidence does not favor a Neogene curvature of at least the eastern half of the arc. However, deep drilling on Meiji Guyot and the Aleutian Abyssal Plain (Scholl and Creager, 1973) raises serious questions about the large displacements postulated by Atwater (1970), Grow and Atwater (1970), and others. If the Cenozoic displacement is measured in hundreds (Scholl and Creager, 1973) instead of thousands of kilometres, it is much easier to envisage the role of the Emperor Ridge in the development of the Aleutian-Kamchatka cusp.

Recent seismicity has been much reduced along the westernmost Aleutian arc, where the Emperor Seamounts adjoin the subduction zone (Hayes and Ewing, 1970) (Fig. 9). This region has also been remarkably free of large earthquakes during the past 75 yr (Sykes, 1971). There is a corresponding gap in late Tertiary andesite volcanoes (Grow and Atwater, 1970), which these authors relate to the fact that plate motion is presently almost tangential to the arc, with little or no subduction. The seismicity gap might be related to this different tectonic setting, provided much of the motion is taken up by creep. In view of the gradual curvature of the arc, the abrupt beginning of the aseismic gap is indeed striking; we suggest the presence of anomalous lithosphere associated with the Emperor Ridge may be a reason for this apparently peculiar behavior. If so, the seismicity of the "gap" will be a permanent feature, contrary to Sykes' (1971) view that it is a high-risk area likely to be hit by a great earthquake in the near future.

Near the eastern end of the Aleutian arc, several seamount chains (here collectively termed the Alaska Seamounts) are not presently associated with significant indentations of the Aleutian Trench or the volcanic arc (Fig. 10). This is not inconsistent with our basic hypothesis (Fig. 3) for two reasons: First, the seamount chains are volumetrically quite small, the onstrike junctions of these chains with the Aleutian Trench being virtually devoid of seamounts. Second, the regime behind the volcanic arc is evidently continental crust (Pratt and others, 1972), not an accreting small ocean basin. These two factors would prevent the Alaska Seamounts from inducing major modifications in the geometry of the Aleutian arc, at least in recent geological time. However, the eastern cusp of the Aleutian arc is not far east of the seamount chains, and it is not inconceivable that now-subducted, more massive extensions of the Alaska Seamount chains influenced the development of this eastern cusp in the past. Alternatively, back-arc extension in the eastern Aleutians and southwest Alaska was prevented by such circumstances as would halt such a phenomenon along

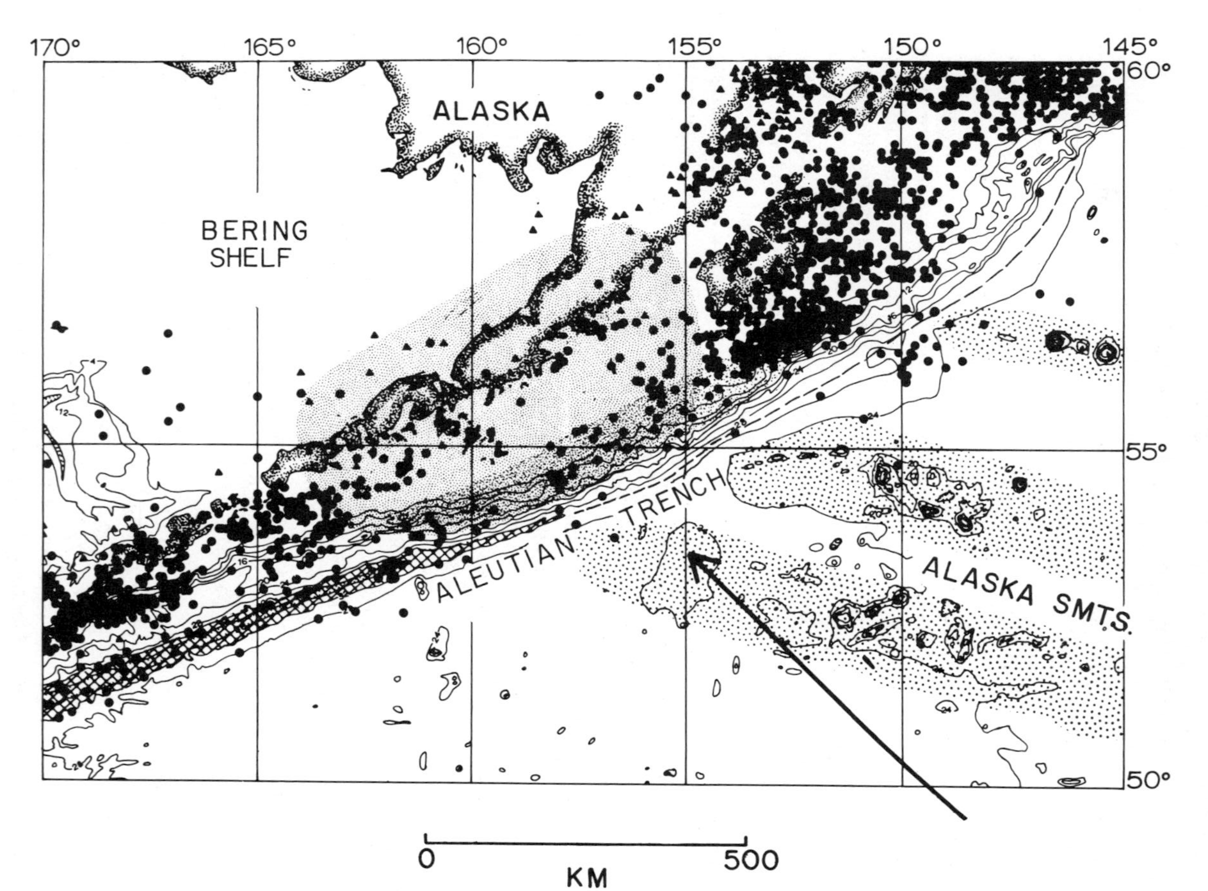

Figure 10. Bathymetry (Anonymous, 1971a, 1971b) and seismicity, intersection of Alaska Seamount chains with Aleutian arc. Irregular stippling shows topographic bulge that replaces mid-slope terrace or valley typically found between trench (dashed) and arc. Symbols as in Figure 4. Much of the "aseismic" zone probably represents the epicentral area of a large (M = 8.3) event of November 10, 1938.

the west coast of the Americas, possibly the relative motion between subduction zone and underlying mantle (Jacobs and others, 1974). Still another possibility is that the northwestward-moving continental margins of Canada and southeastern Alaska acted as an "aseismic ridge" to pin the Aleutian arc in the area of the cusp.

The age-dating and trends of the chains (Turner and others, 1973) suggest that they are Neogene hot-spot trails.

Although there is presently no cusp where the Alaska Seamounts intersect the Aleutian Trench, the shape of the sea floor in the arc-trench gap between the Aleutian Ridge and the trench axis is different in the seamount-arc intersection than it is on either side (Fig. 10). In the intersection area, the shallow shelf bulges southward toward the trench; the platform separating the deep trench floor from the island shelf west of 162°W disappears where the seamounts come in. East of the seamount intersection, the edge of the continental shelf again retreats from the trench axis. Although the 10^5 km^3 extra volume could represent sheared-off seamounts, sheared-off pelagic sediments, substantial local terrigenous sediments (Scholl and Creager, 1973), and differential downwarping of the bulge area must be important also. Further, the bulge may be cored by a postulated Cretaceous subduction zone that extended seaward of the present Alaska Peninsula, through the Kodiak and Shumagin Islands (Burk, 1965, 1972; Moore, 1973).

There is a prominent region of reduced seismicity (1961–1971) where the Alaska Seamount chains intersect the Aleutian arc (Fig. 10). The relatively aseismic character of this region is still apparent when all shallow events of magnitude >7.0 since 1900 and all aftershocks since about 1930 are totalled (Sykes, 1971). Unlike the Emperor-Aleutian intersection, this one is not associated with any prominent geometric irregularity in the trace of the arc. This reduces the number of explanations for the reduced seismicity: Either the aseismic zone is an artifact of insufficient observation time, or the subducted seamount chains and the lithosphere below them are somehow responsible for the reduced seismicity. The first explanation is favored by Shakal and Willis (1972), who concluded that there is an 80 percent probability of an $M_S \geq 8$ earthquake occurring between 155°W and 167°W before 1980. Such an event would undoubtedly be followed by an aftershock sequence which would remove some of the aseismic character apparently associated with the downgoing seamount chains. However, there is no certainty that a 42-yr observation period gives accurate recurrence probabilities for large events (Shakal and Willis, 1972). We suggest, but without strong conviction, that the reduced seismicity associated with this region in the period 1961–1971 is a permanent feature. The small southwestward shift of the aseismic region with respect to the seamount-trench intersection can be explained in part as an artifact of projection: The extensions of the seamount chains in the downthrust slab appear to bend more east-west when projected on a map of the Earth's surface. The heavy concentration of events south of Kodiak Island is not accounted for this way, however, and any extension of the northern seamount chain along the downgoing Pacific plate must pass through this seismic zone unless the events have occurred in the Americas plate. One or two more decades of observation should be enough to determine whether the proposed aseismic region is a permanent feature. Although the historical record prior to 1930 is certainly inadequate, earlier records of large magnitude events are of some use. Sykes (1971) has relocated earthquakes larger than magnitude 7 since 1900 and most aftershocks from main shocks after about 1930 using earthquake data since 1900. The total number of main shocks and aftershocks in the interval 155°W to 162°W is still less than half that of adjacent segments of the arc. This might be taken as good evidence for permanently reduced seismicity were it not for the fact that the one great earthquake occurring in the region was the largest seismic event in the Aleutian arc (8.7), and it occurred in this area in 1938. Our proposal that seismic activity is reduced is only weakly supported here, at least for great earthquakes and their aftershock sequences.

Caribbean and Scotia Arcs

In geometry and tectonic setting, the Scotia and Caribbean (West Indies or Antilles) arcs are strikingly similar. This similarity even extends to apparent modifications of the northern plate boundaries by aseismic ridges on the overridden plate. The massive Bahama Ridge extends southeast into the plate boundary offset northeast of Hispaniola (Fig. 11), while the South Georgia Rise lies northeast of a similar boundary complication associated with South Georgia Island (Fig. 12). There are several significant differences between the Scotia and Caribbean areas, however, and as a consequence, slightly different models are suggested to attribute these complications in the plate boundaries to aseismic ridges on the downgoing plate (Figs. 13, 14).

The northern boundary of the Caribbean plate follows the northern edge of the eastern Cayman Trough (Holcombe and others, 1974) and then bends southeast along the northeast coast of Hispaniola to join with the Puerto Rico Trench (Fig. 11). Bracey and Vogt (1970) suggested that northeast Hispaniola is behaving as a "miniarc," with a larger component of underthrusting than under Puerto Rico. The plate boundary in the Hispaniola–Puerto Rico area is evidently not a simple one, however. South of eastern Hispaniola and Puerto Rico, an additional north-dipping thrust zone seems to isolate these islands from both the Americas and Caribbean plates (Matthews and Holcombe, 1974). This thrust zone, marked by the Muertos Trough, appears to be offset right laterally along northeast-trending lineaments (faults?) where it approaches Hispaniola (Matthews and Holcombe, 1974). The structure then continues on a more westerly strike in the form of the Enriquillo–Cul-de-Sac Valley of Hispaniola, and can be followed into the Cayman Trough (Fig. 11). Toward the east the Muertos Trough structure may continue northeast into the Anegada Trough, a probable transcurrent fault zone. Another probable active or recently active fracture trough lies between Jamaica and the southwestern tip of Hispaniola (T. Holcombe, 1974, personal commun.).

We suggest that these complexities in the plate boundary represent a splintering of the northern Caribbean plate as it began to override the massive Bahama Ridge (Fig. 13). Molnar and Sykes (1969) also suggested that northern Hispaniola may be a separate splinter from the main Caribbean plate. The Bahama Ridge appears to narrow toward the southeast; some of this narrowing may be due to partial subduction of the southwestern margin of the ridge under Cuba and Hispaniola. The surviving Bahama Ridge can be traced southeast past Navidad Bank to a seamount at 67.5°W, equidistant from the tips of Puerto

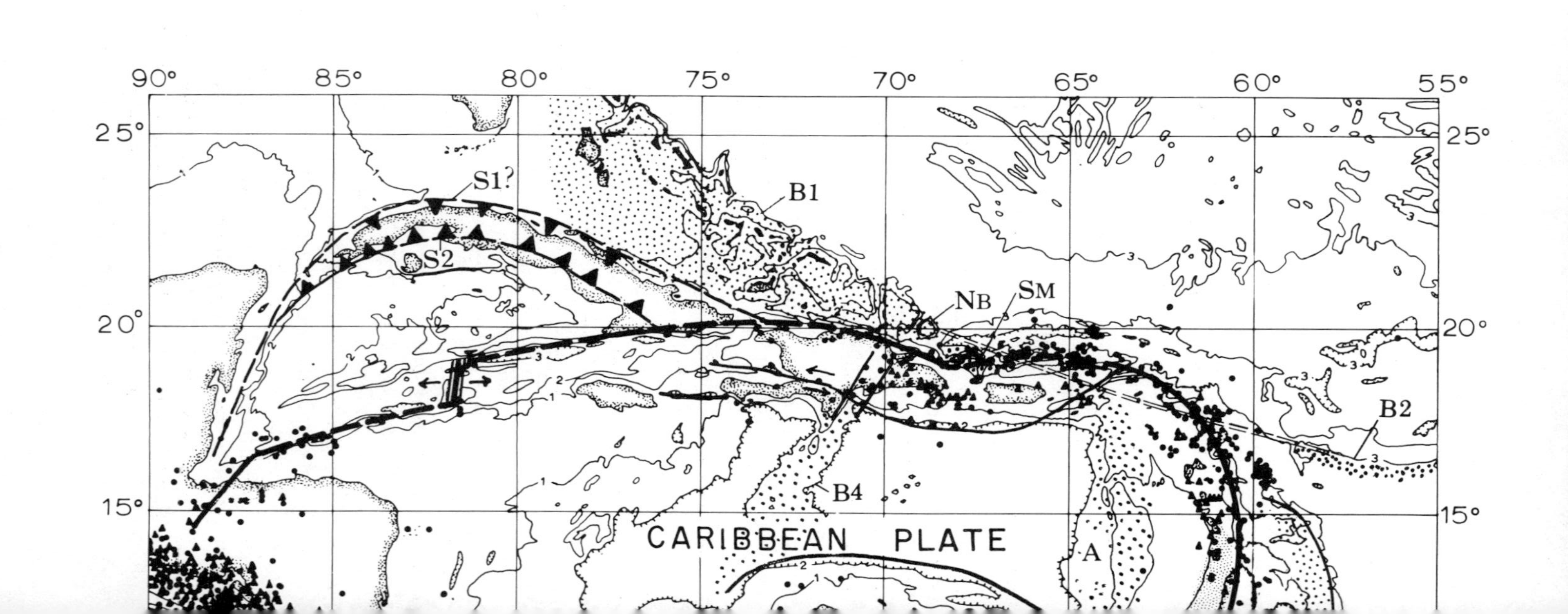

90°
85°
80°
75°
70°
65°
60°
55°
25°
20°
15°
S1?
S2
B1
NB
SM
B2
B4
A
CARIBBEAN PLATE

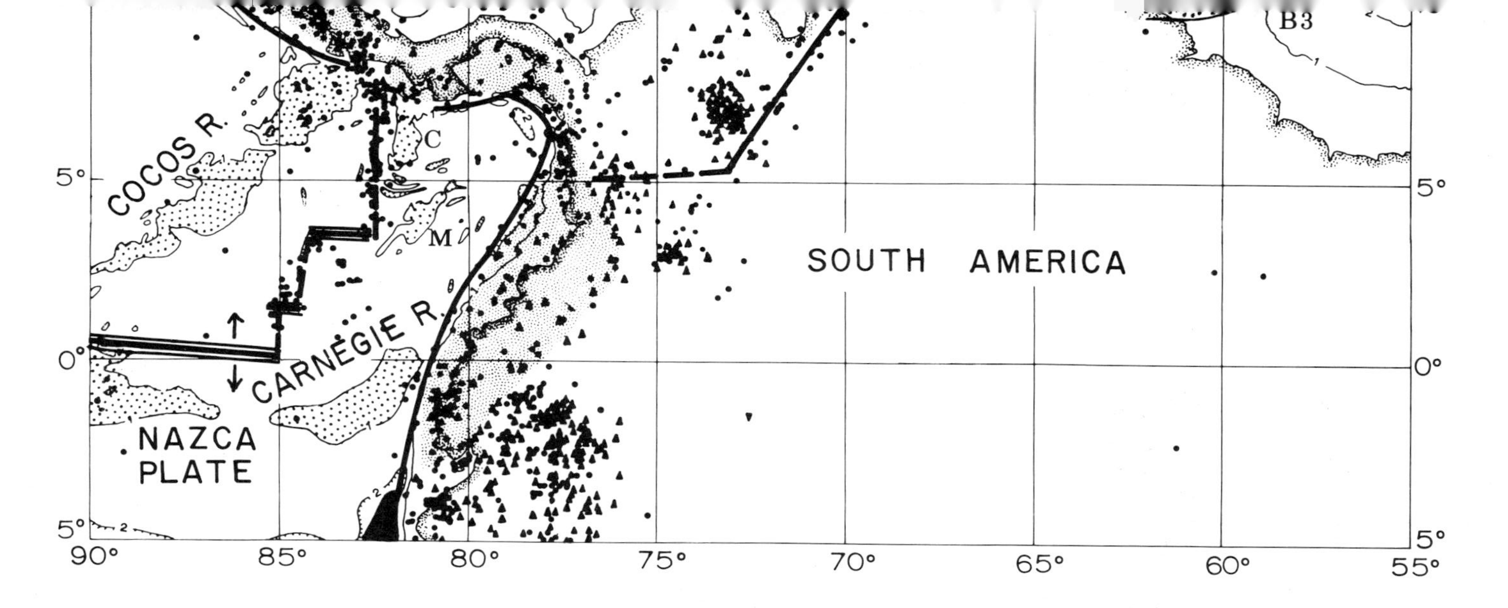

Figure 11. Bathymetry (Anonymous, 1972a) contoured at 1,000 fm (1,850 m) intervals, plate boundaries, and seismicity in Caribbean area. Heavy lines show present plate boundaries. Arrows show spreading direction. A, Aves Ridge; B1, Bahama Platform and Ridge; B2, Barracuda Ridge (Peter and others, 1974); B3, Barbados arc, or ridge; B4, Beata Rise; NB, Navidad Bank; SM, most southeastern seamount of the Bahama Ridge; double-dashed line, hypothesized former connection between Barracuda and Bahama Ridges; S2, north-dipping fossil subduction zone along south coat of Cuba postulated by Malfait and Dinkelman (1972); S1?, possible earlier south-dipping subduction zone suggested by northward convexity of Cuban arc; C, Coiba Ridge; M, Malpelo Ridge.

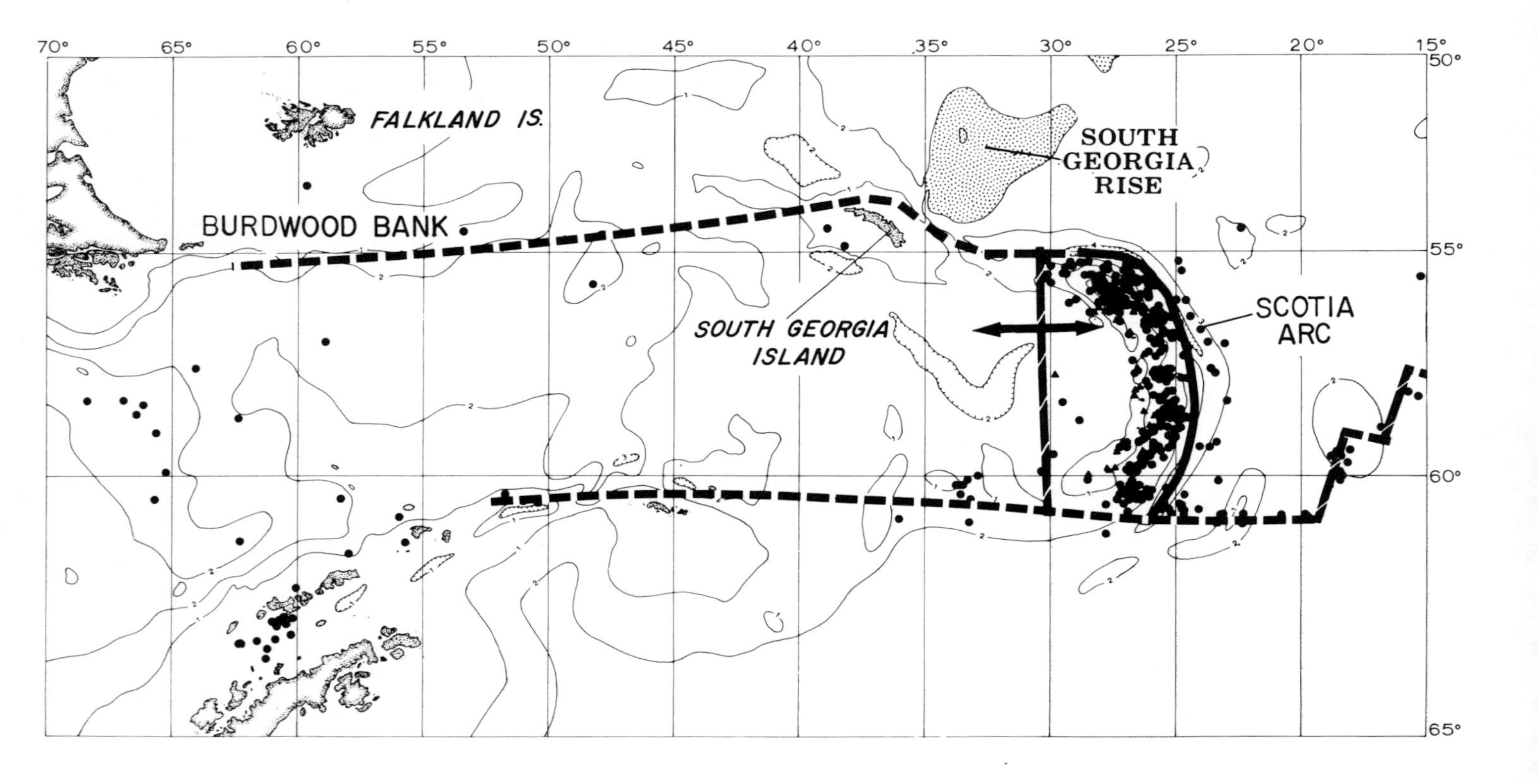

Figure 12. Bathymetry (Anonymous, 1972a) contoured at 1,000 fm (1,850 m) intervals, plate boundaries, and seismicity (Anonymous, 1972b) of the Scotia arc. Heavy lines show present plate boundaries. Arrows show spreading direction.

Rico and Hispaniola (Bowin, 1975) (Fig. 11). Much farther east, the Barracuda Ridge (also called a fracture zone) lies on strike and parallel to the eastern Bahama Ridge. The Barracuda Ridge is a basement high about 50 km wide and 2 km above the surrounding Atlantic Basin (Peter and others, 1974). We suggest the Barracuda and Bahama ridges represent a single, formerly continuous feature partly destroyed by subduction (Fig. 13).

The northeastern edge of the Bahama Ridge is a steep escarpment that, to judge from its trend, may well be a transform fault indicating the Mesozoic trace of motion between the Africa and North America plates (Pitman and Talwani, 1972; Le Pichon and Fox, 1971; Vogt, 1973b). When Africa and North America are reconstructed, the Bahama Ridge (or platform) severely overlaps continental Africa. The ridge, therefore, represents an "excrescence," formed on oceanic lithosphere of the North America plate in the Jurassic or Cretaceous, some time after the continents had moved apart (Dietz and others, 1970). Perhaps the ridge began as a hot spot like Iceland or the Azores. This would explain the shallow initial depths required to form carbonate reefs; the igneous crust would also be thickened. The eastern end of this ridge (the Barracuda Ridge) was perhaps too deep to nucleate carbonate reefs.

Whatever its role in plate tectonics, the massive, quasi-continental crust of the Bahama Ridge would resist subduction and complicate the plate boundary in the Hispaniola–Puerto Rico area. The model proposed to explain these complications (Fig. 13) does not involve back-arc spreading (Fig. 3) because geologically youthful extension has not been proved to occur in the Caribbean (this is not true for the Scotia arc, however, as we shall discuss later). Although the Aves and Beata Ridges may well be remnant arcs, the basin between them or between the Aves and Antilles would have had to form *after* the first contact between the arc and the Bahama-Barracuda Ridge for the mechanism of Figure 14 to operate. The model for the Bahama Ridge–Caribbean plate interaction (Fig. 13) is undoubtedly too simple, but it does suggest that left-lateral motion should be occurring along the Enriquillo-Muertos lineament, as Hispaniola is dragged against the Bahama Ridge.

At current rates of plate motion, the aseismic ridge began to collide with the arc in about mid-Tertiary times, and the collision is apparently still in progress. A large lateral displacement such as suggested in Figure 13 could actually be distributed among many lesser faults. A large displacement was originally proposed by Hess and Maxwell (1953) but remains unproven (Bowin, 1975; Malfait and Dinkelman, 1972; Meyerhoff and Meyerhoff, 1972). However, large horizontal displacements are difficult to prove, especially on Hispaniola where field work is tedious and so far highly inadequate (Khudoley and Meyerhoff, 1971). A large Tertiary displacement between northern Hispaniola and the main Caribbean plate is also implied if that island was once attached to Cuba (Malfait and Dinkelman, 1972), for the motion required to reconstruct those islands into a single landmass is much less than the total opening of the Cayman Trough (Holcombe and others, 1974). Whatever the strike-slip displacement, the Enriquillo and other grabens of Hispaniola have subsided and collected at least several kilometres of sediments since middle Tertiary time. The Miocene section is particularly thick; unlike Miocene sections elsewhere in the Greater

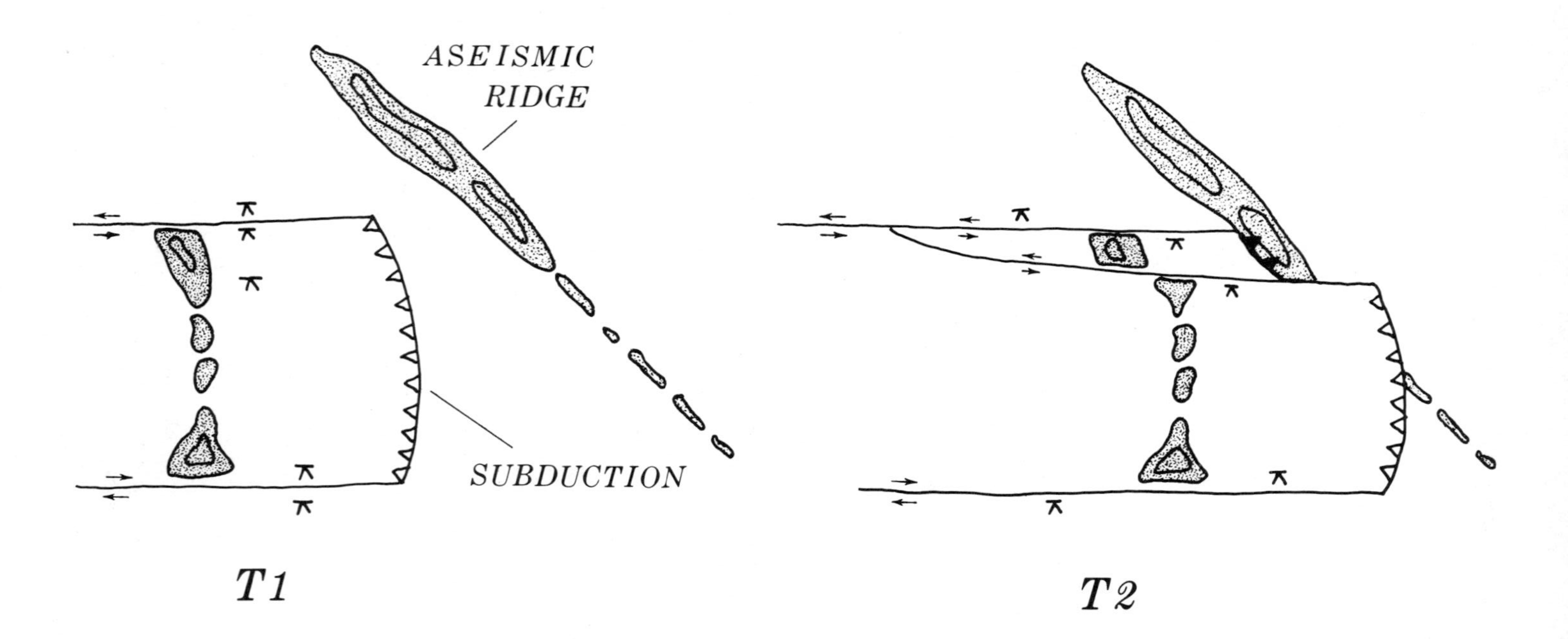

Figure 13. Proposed interaction of aseismic ridge (Bahama-Barracuda Ridge?) with island arc if inter-arc extension does *not* occur. Symbols and remnant arc are shown to illustrate relative offsets between overridden plate (Atlantic), main overriding plate (Caribbean?), and small northern subplate (northern Hispaniola and Puerto Rico?). Compare with Figure 11, but note that inter-arc extension might also have occurred between times T1 and T2 in the Caribbean, so that the actual evolution of the area would involve the relationship of Figure 14.

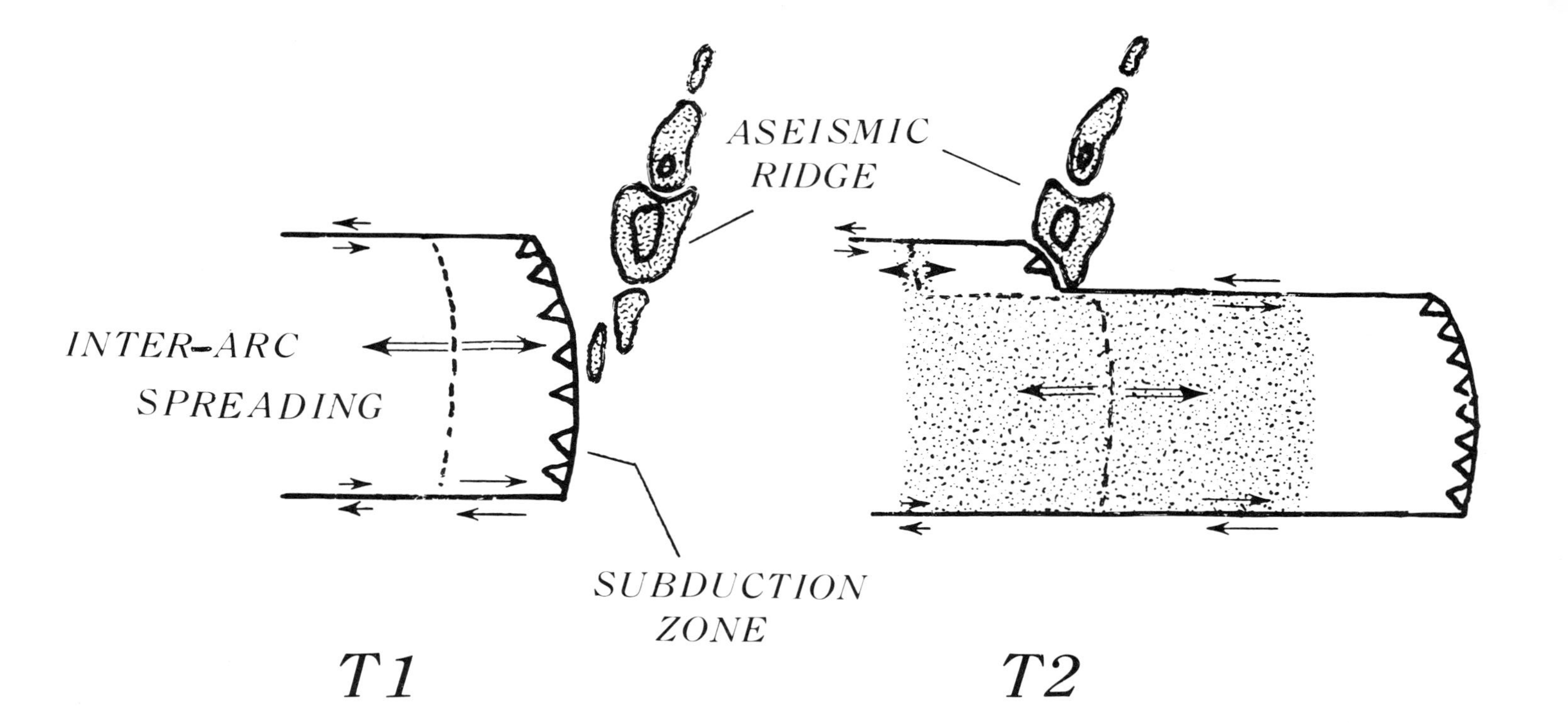

Figure 14. Proposed interaction of aseismic ridge with an arc bounded by major transforms, based on Figure 3. Inter-arc spreading is reduced or eliminated behind massive portion of aseismic ridge, possibly explaining the position of South Georgia with respect to the Scotia arc (Fig. 12).

Antilles, those of Hispaniola are strongly to moderately deformed, as are Pliocene beds (Khudoley and Meyerhoff, 1971). We attribute this unusual deformation to the influence of the Bahama Ridge. Thrusting, perhaps even northward subduction, of the main Caribbean plate under a northern subplate composed of Hispaniola and Puerto Rico (Matthews and Holcombe, 1974) may come about as the Bahama Ridge presses this fragment southward against the crust of the Caribbean Basin. The Muertos Trough underthrusting apparently began in middle to late Tertiary times, in agreement with the supposed time of first collision of the Caribbean plate with the Bahama Ridge.

Interestingly, the Muertos Trough and a similar zone of deformation along the south edge of the Caribbean plate (Fig. 11) are also arcuate, both being convex toward the interior of the plate. T. Holcombe (1974, personal commun.) suggested that these ridges have acted as rigid, buoyant "beams" that resisted compression of the plate as a result of late Tertiary convergence between North and South America (Ladd, 1974).

We note the possible significance of two quite different arclike structures on the adjacent Americas plate. The first is the Barbados Ridge, which from geophysical and geological data has been proposed to be a thick sediment pile tectonically shaped by the downthrusting Atlantic lithosphere (Ewing and others, 1970; Bunce and others, 1970). However, the ridge could also be a pre-early Miocene island arc, as implied by Meyerhoff and Meyerhoff (1972). The same authors point out that the Barbados Ridge terminates east of Marie Galante Island, the southern end of the Limestone Caribbees. These islands evidently represent a pre-early Miocene island arc. The once continuous(?) Barracuda-Bahama Ridge (Fig. 11) also intersects the Antilles near the southern end of the Limestone Caribbees. These curious relationships and other data suggest that the Limestone Caribbees and the Barbados Ridge may have formed a continuous subduction zone until the late Eocene, when the northern part of the arc first collided with the Bahama-Barracuda Ridge. The southern section of the arc became detached along an arc-arc transform at 15°N and migrated farther east, perhaps by back-arc spreading (Fig. 14). A subduction zone then developed during the Oligocene inside the Caribbean plate, thereby stranding the Barbados Ridge on the Americas plate. This new subduction zone consumed most of the crust once lying between it and the Barbados-Limestone arc, with the present arc becoming attached to the earlier Limestone arc perhaps in the late Oligocene or early Miocene. Thus, it may be the difficulty experienced by the east Caribbean subduction zone in consuming the semicontinental Bahama Ridge that has slowed and complicated the eastward march of this plate. The subduction of the Barbados Ridge, although a different class of feature than other aseismic ridges, seems to be accompanied by relatively reduced seismicity (Fig. 11).

The Cuban "arc" is a second example of an arcuate tectonic belt presently attached to the Americas plate. Malfait and Dinkelman (1972) have speculated that Caribbean crust was actively underthrusting the Cuban arc from the south from Cretaceous through Eocene times. The arc continued east through Hispaniola and Puerto Rico, but underthrusting was from the Atlantic side there. Activity along the Cuban subduction zone came to an end in the Eocene; a new plate boundary formed along the Cayman Trough, thus transferring the Cuban arc to the Americas plate. Two observations make the Cuban arc interesting in the context of this paper: (1) The arc abuts the Bahama–south Florida platform on the east and north and the Yucatan shelf on the west; and (2) contrary to almost all known island arcs, the trench lay on the *concave* side of the arc, if Malfait and Dinkelman are right. This line of reasoning suggests that as an early, as yet undiscovered *north-facing* Cuban arc migrated north in Cretaceous times, consuming oceanic crust of Jurassic or Early Cretaceous age, it came in contact with the continental or quasi-continental platforms. This caused an arc polarity reversal (Fig. 2) according to the definition of Karig (1972), with the old Caribbean lithosphere now underthrusting Cuba from the south. Alternatively, there was only a single north-facing arc, which became inactive when most of the ocean crust north of it was consumed.

We now consider the Scotia arc and its tectonic setting (Dalziel and Elliot, 1973) (Figs. 12, 14). The plate boundaries in the Scotia Sea area are not well defined except near the volcanic arc and South Sandwich Trench. As pointed out earlier, there are some striking parallels with the Caribbean arc; in particular, South Georgia Island occupies a bulge in the supposed plate boundary, similar to Hispaniola on the Caribbean plate. An aseismic oceanic ridge, the South Georgia Rise, occupies a position on the Americas plate northeast of South Georgia, comparable to that of the Bahama Ridge northeast of Hispaniola (Fig. 11). The South Georgia Rise is a more subdued bulge than its northern counterpart, at least in part due to the absence of carbonate reefs on its crest. The rise is a broad, complex feature, only poorly charted and of unknown crustal structure. Minimum depths are less than 2 km, versus 4 to 5 km for the more typical ocean basins surrounding it. A Late Cretaceous age for the rise is obtained by extrapolating crustal isochrons associated with the southern Mid-Atlantic Ridge to the east.

The Scotia area differs from the Caribbean in that the South Sandwich arc is connected to the Mid-Atlantic Ridge by a series of spreading centers and transform faults extending eastward from the south end of the trench. Also, there appear to be two active spreading axes in the Scotia Sea, one in the Drake Passage between South America and Antarctica, and a north-south axis at 33°W, somewhat west of the volcanic arc (Barker, 1972; Barker and Griffiths, 1972). The spreading from these ridges dates from at least 20 m.y. B.P. and about 8 m.y. B.P., respectively. The eastern ridge is of interest in the context of this paper, for it suggests back-arc spreading is taking place. The type of interaction of such a system with an aseismic ridge on the overridden plate would, therefore, be along the lines of Figure 14. Until about 8 m.y. B.P., South Georgia was the northern end of a proto–South Sandwich arc. When this system bumped into the South Georgia Rise, a new spreading center developed around 36° to 38°W, detaching the South Georgia platform and leaving behind a subdued remnant arc at around 36° to 38°W. Subsequent displacement of the South Georgia Rise with respect to South Georgia has been relatively slight, as suggested by the subdued seismicity (Fig. 12).

It might be argued that the South Georgia platform is not really an old island arc, despite its curvature and relation to the South Sandwich arc. According to Dalziel and Elliot (1973), metamorphic and sedimentary rocks as old as Paleozoic may occur there, as well as Cretaceous and younger ones. Those authors interpret South Georgia as formerly a part of the Pacific hinterland of the Andes. The tectonic curvature of South Georgia may date from the Late Cretaceous or early Tertiary (Dalziel and Elliot, 1973). However, the dating is largely based only on lithologic similarities to other formations. Even if South Georgia is a fragment of South America, it was evidently carried east through a transform plate boundary between it and Burdwood Bank to the north. There is no reason why it could not have been part of an eastward migrating arc, just as the Japanese "arc" has migrated east long after its earliest rocks had been emplaced there (Uyeda and Miyashiro, 1974). Dikes of Tertiary age in South Georgia (Dalziel and Elliot, 1973) may then have been emplaced when the island was part of the active arc. Conversely, the fact that the emergent South Sandwich volcanics are not known to predate 4 m.y. B.P. (Baker, 1968) does not rule out substantially older rocks at depth in the arc ridge.

Tehuantepec, Cocos-Carnegie, and Nazca Ridges: Subduction of Aseismic Ridges without Inter-arc Spreading

We have discussed how aseismic ridges on the overridden plate might form notches or cusps in the trace of the subduction zone if back-arc extension can occur. The Tehuantepec, Cocos-Carnegie, and Nazca Ridges, as well as the Alaska Seamount chains mentioned earlier, represent cases where there are no marginal basins behind the arcs. Nevertheless, there are "central valleys" along parts of the Central and South American subduction zones. Further, marginal basins may have bordered the west coasts of the Americas in the past, as they apparently did in the southern Andes (Dalziel and others, 1974). If tensional in origin, the central valleys may represent modest or incipient phases of back-arc extension insufficient to produce new ocean crust and marginal basins (Karig, 1972). To a limited degree, therefore, Andean-type subduction zones could become gently notched when aseismic ridges are subducted. Indeed, the Tehuantepec Ridge trends into a gentle but unmistakable notch in the Middle America Trench. The Cocos, Coiba, and Carnegie Ridges do not trend into individual cusps, but this complex of ridges does coincide with a larger order, eastward-pointing notch—the Bay of Panama—between the North and South American west coasts. The Nazca Ridge does not trend into a notch; although it is not far north of the bend between the Peru and Chile Trenches, this is probably fortuitous, since the ridge-trench intersection is presently moving south. Farther south, the Juan Fernandez Island and seamount chain trends into another gentle notch in the Chile Trench (Fig. 1). The association of cusps with aseismic ridges from the eastern Aleutian Trench (Fig. 10) to the central (Fig. 11) and South American (Fig. 1) subduction zones is thus only partial, and perhaps fortuitous. However, in these areas a good correlation was not to be expected from our model (Fig. 3).

Because there should be no major geometric modification of an Andean subduction zone by a downgoing ridge, it becomes possible in those cases to investigate whether subducted ridges modify the seismicity, volcanism, or other properties apart from possible "geometric" effects. Thus, the aseismic western end of the Aleutian arc (Fig. 9) or the aseismic Yap arc (Fig. 5) could simply reflect differences in subduction rate or in the component of underthrusting rather than differences in the properties of the plate containing the ridge. The meaning of the relatively aseismic areas in those places is, therefore, ambiguous. We recall, however, that there has been a prominent aseismic region (1961–1971) where the Alaska Seamounts intersect the "Andean"-type eastern Aleutian arc (Fig. 10). If the depressed seismicity is an artifact of the observation interval, the chance of another such association between a downgoing ridge and reduced seismicity would be small.

To examine the seismicity of the subduction zones in the vicinity of the Tehuantepec (Fig. 1), Cocos (Fig. 11), and Nazca (Fig. 1) Ridges, we first divided the Benioff zone into corridors ½° (55 km) wide (Figs. 15 to 17). These corridors extend down the slab parallel to the projected continuation of the subducted ridges. Events (1964–1971) were divided by depth and magnitude and composite histograms constructed. The histograms are referred to the Earth's surface along the trench in order to facilitate comparison with the surviving portion of the ridges. Quaternary andesite volcanoes are also shown in simplified fashion. In calculating the extensions of the ridges in the descending slab, we used the distribution of earthquake foci, in sections perpendicular to the arcs, to define the slabs.

The seismicity profile (Fig. 15) suggests that there is relatively reduced activity associated with the subducted extension of the Tehuantepec Ridge. Actually, the area of reduced seismicity is broader than the ridge itself. The effect is more pronounced for higher magnitude events, although less valid statistically, and it seems to affect all depth intervals. However, a deepening of deepest events is also evident south of the Tehuantepec Ridge—middle America subduction zone intersection. Other effects associated with the entry of the Tehuantepec Ridge into the trench have also been noted by Menard and Fisher (1958) and Truchan and Larson (1973): There is a gap in the line of andesite volcanoes, and the volcanoes are much farther inland northwest of the intersection than they are to the southeast (Fig. 1); the trench is deeper and V-shaped in the south, but flat and shallower in the north (this could reflect sediment filling in the north, and damming by the Tehuantepec Ridge); the oceanic crust under the trench is also thought to be thicker southeast of the intersection (Shor and Fisher, 1961). It thus appears that the Tehuantepec Ridge is associated with a cusp in the trench system, as well as prominent disconti-

→

Figure 15. Seismicity (1964–1971) (Anonymous, 1972b), Quaternary volcanoes (top) (Dengo, 1973), and bathymetry profiles where Tehuantepec Ridge intersects Middle America Trench. Only events located by 10 or more recording stations occurring from January 1964 through August 1971 were used. Focal depths (top) and magnitudes (middle) were summed along ½° (55 km) wide corridors along the downthrust slab in the projected continuation of the aseismic ridge in the slab. Bathymetric profile shows trench axis (bottom); crest of aseismic ridge seaward of trench indicated by heavy bar. Triangles at top denote number of active volcanoes at each site, as estimated from Dengo (1973).

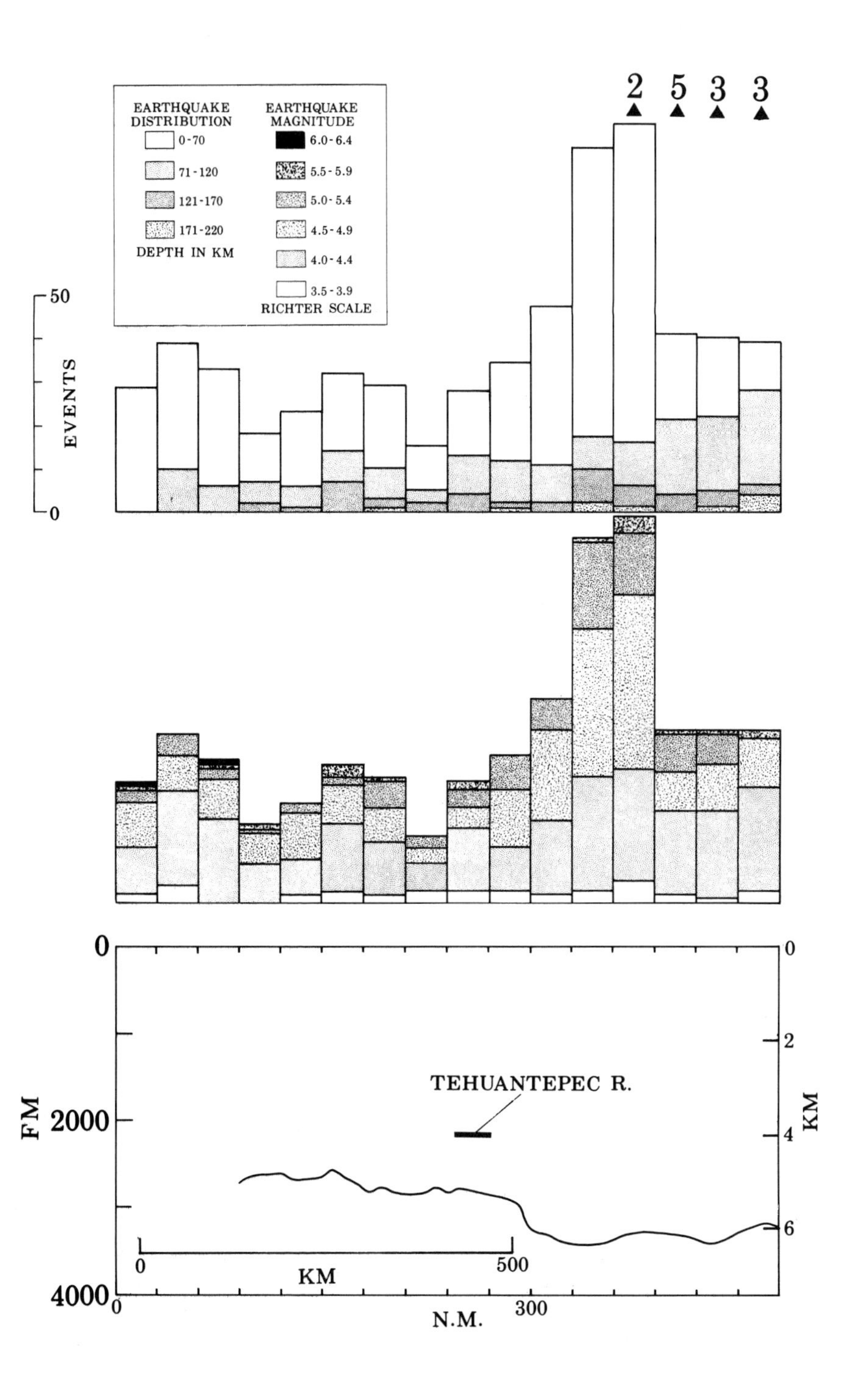

EARTHQUAKE DISTRIBUTION
0-70
71-120
121-170
171-220
DEPTH IN KM
EARTHQUAKE MAGNITUDE
6.0-6.4
5.5-5.9
5.0-5.4
4.5-4.9
4.0-4.4
3.5-3.9
RICHTER SCALE
2
5
3
3
50
EVENTS
0
0
FM
2000
4000
0
2
4
6
KM
TEHUANTEPEC R.
0
500
KM
0
300
N.M.

nuities in volcanism and seismicity. We relate these effects to the subduction of the Tehuantepec Ridge and to properties of the lithosphere containing the ridge. The approximate parallelism of the ridge and the Americas-Cocos relative motion (Sclater and others, 1971) means that the Tehuantepec Ridge has trended into the Gulf of Tehuantepec area for some time.

The Carnegie and Cocos Ridges, and several smaller ridges such as the Coiba and Malpelo Ridges, seem to be parts of a single feature. As we pointed out before, the Cocos and Carnegie Ridges do not trend into local cusps, although the whole complex of ridges forms a larger order eastward embayment in the boundary of the Americas and Caribbean plates, an observation that may or may not be fortuitous. At least, the entire region between and including the Cocos and Carnegie Ridges is relatively elevated and young and therefore not as readily subductible as adjacent lithosphere. Van Andel and others (1971) suggested the Coiba-Malpelo Ridge was once on the Cocos plate; it plugged up the subduction zone and caused the development of the Panama fracture zone to the west. This transferred the ridge to the Nazca plate. A further switch of the northeastern end of the Cocos Ridge to the Nazca plate is in progress, according to those authors. The evidence available today does not favor the mechanism of van Andel and others for the position of the Cocos and Carnegie Ridges; however, such a mechanism would still suppoprt our basic thesis that aseismic ridges are not readily subducted.

Seismicity and volcanism seem to be affected where the Cocos Ridge is being subducted (Fig. 16). Quaternary volcanoes become less frequent and disappear; earthquake seismicity also decreases toward the ridge-trench intersection, especially intermediate and deep events, and for magnitudes in the range 4.5 to 5.4, as in the case of the Tehuantepec Ridge (Fig. 15), the effect is broader than the prominent, bathymetrically defined Cocos Ridge. Until crustal isochrons on the Cocos plate are better known, we cannot exclude decreasing plate age (hence thickness) as a possible explanation for the decrease in volcanism and seismicity. Further, an incipient plate boundary may be developing near 85°W, west of the Panama fracture zone (van Andel and others, 1971); thus the Cocos Ridge may be forced into the mantle at a slightly lower rate than the sea floor to the west. However, spreading rates on the Costa Rica rift zone are not measurably less than on the Galapagos Ridge just to the west, as they should be if such differential motion were occurring.

The Nazca Ridge (Herron, 1972; Morgan, 1973) and the Juan Fernandez Islands and seamounts (Figs. 1, 17) seem to be associated with discontinuities in the Peru-Chile subduction zone, perhaps because there is no "back-arc extension" in the Andes. There is a slight reduction of earthquake seismicity (Fig. 17), again over a broader area than the Nazca Ridge itself. A number of fundamental geological discontinuities mark the Nazca-Andean intersection. The intersection marks (1) the northern limit of the Quaternary Chilean andesite volcanoes (Fig. 17); (2) the northern limit of the longitudinal or central valley west of the volcanic chain; (3) the northern limit of the flat-floored, elevated basin called the Altiplano; (4) the southern limit of the deep Peruvian earthquakes (Ocola, 1966); (5) the northern limit of "deep" intermediate depth Chilean seismicity; (6) the major structural discontinuity called the Pisco De-

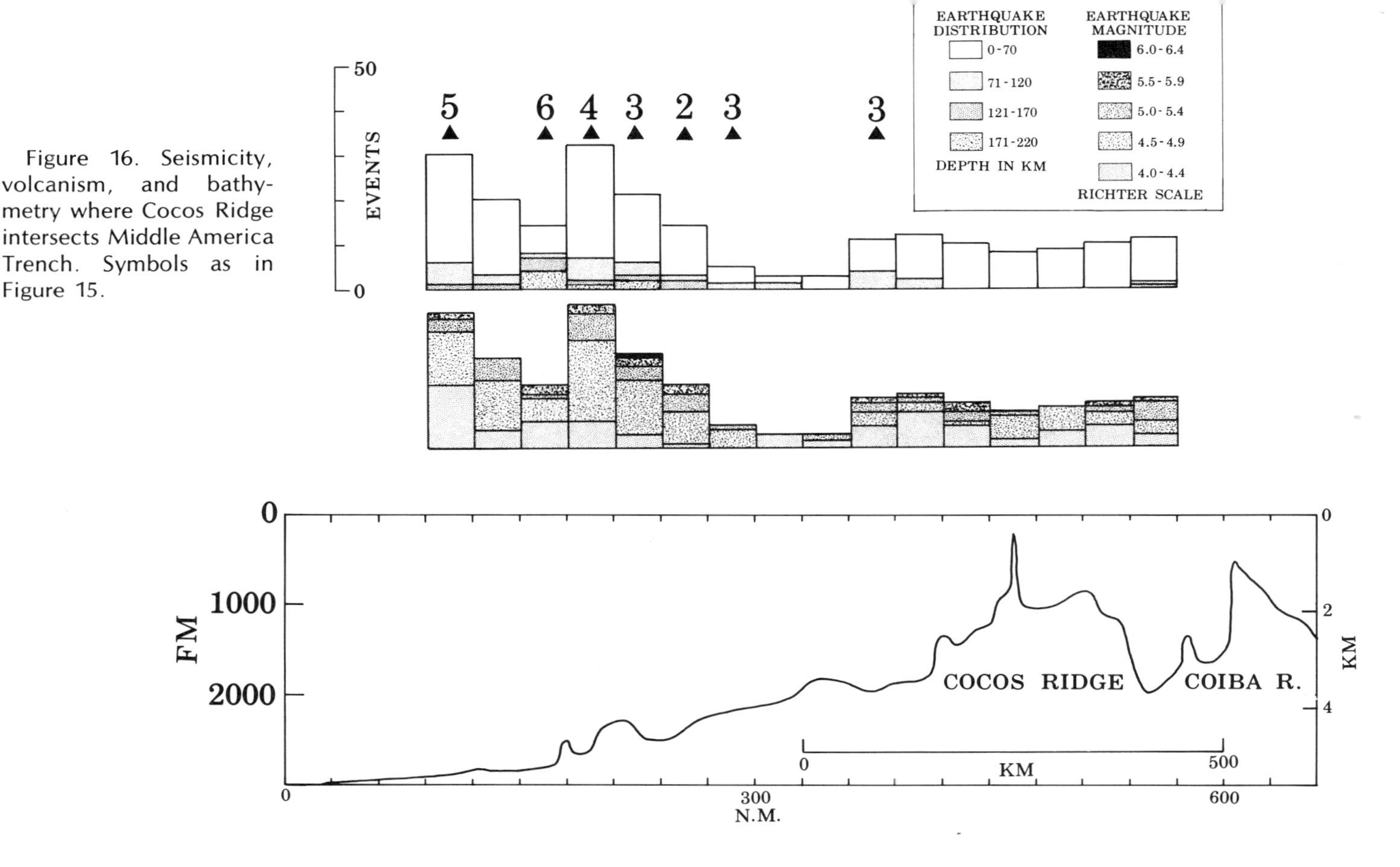

Figure 16. Seismicity, volcanism, and bathymetry where Cocos Ridge intersects Middle America Trench. Symbols as in Figure 15.

Figure 17. Seismicity, volcanism, and bathymetry profiles where Nazca Ridge intersects Peru-Chile Trench. Symbols as in Figure 15. Volcanoes from Pichler and Zeil (1969), Plafker and Savage (1970), Casertano (1963), and Operational Navigation Charts, courtesy of the Aeronautical Information Center of the U.S. Air Force.

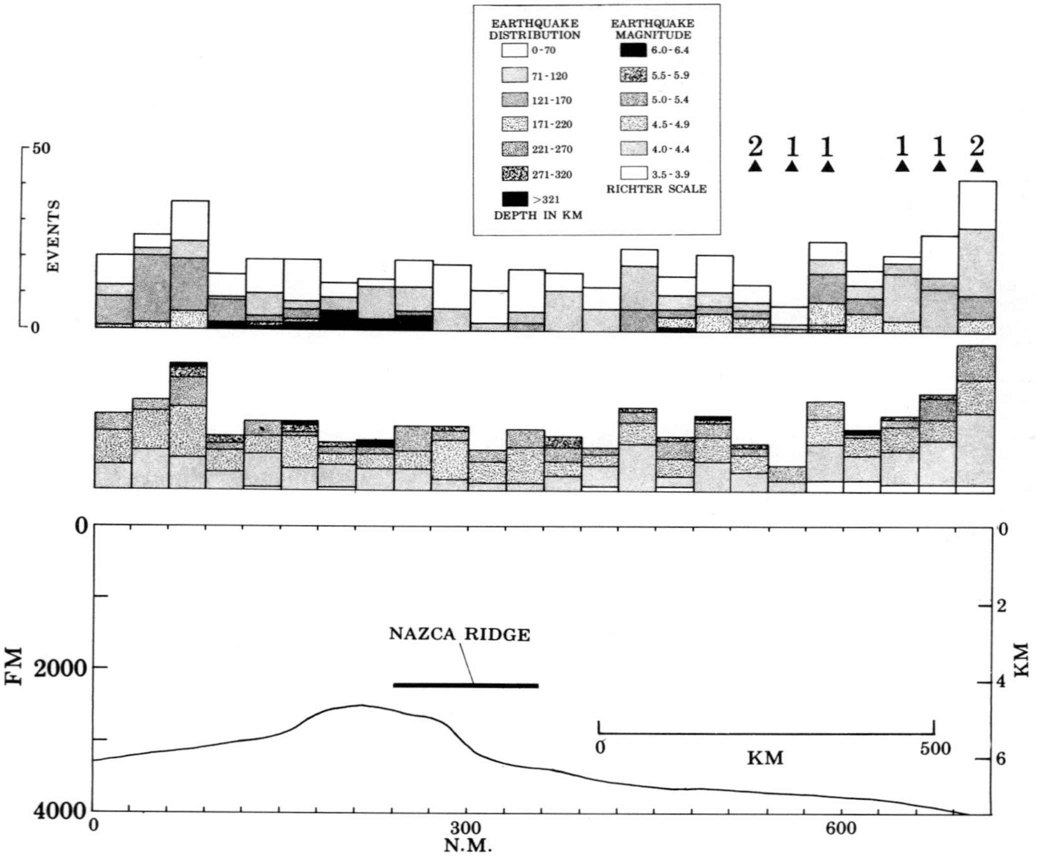

flection; and (7) other changes as noted by Sillitoe (1974), who refers to this discontinuity as "tectonic boundary No. 4." Some of the above coincidences may be fortuitous, but it is hard to believe they all are. The Nazca Ridge subduction may be responsible for some of these effects.

Our final example of a subducted aseismic ridge is the Juan Fernandez Islands and seamounts, called the Chile fracture zone by Sillitoe (1974). This feature is about 50 km wide and seems to be a chain of isolated peaks rather than a continuous ridge like the Nazca, Cocos, and Carnegie Ridges. The adjacent sea floor is mid-Tertiary in age (Herron, 1972). The Juan Fernandez chain enters a gentle cusp in the Chile Trench and thus resembles the Tehuantepec Ridge. Like the latter and the Nazca Ridge, there is a volcanic discontinuity in the intersection area; the south Chilean Quaternary andesite volcanoes and the associated central valley have their northern limit here. Seismic activity and Andean summit elevation decrease sharply south of the Juan Fernandez–Chile Trench intersection. Other changes are discussed by Sillitoe (1974) who calls this "tectonic boundary No. 13."

Conclusion

A worldwide inspection of aseismic ridges being subducted (Fig. 1) reveals too many associated cusps and other irregularities to support the belief that these ridges do not affect the geometry of subduction zones. Perhaps some of the coincidences we have adduced are fortuitous; it is hard to believe that many are. One reasonable explanation is that the extra buoyancy imparted to the downgoing plate locally arrests the "seaward" migration of the frontal arc; that is, it slows down back-arc extension (Fig. 3). If island arcs are not free to migrate toward the consumed plate as proposed by Karig and other authors, the coincidence between cusps and aseismic ridges would be extremely difficult to comprehend. (In a few cases the geometry of the subduction zone could have been modified by splintering of the consuming [Figs. 11, 13] or consumed plate [van Andel and others, 1971; Karig and others, 1973]). Where no back-arc extension occurs, cusps should not readily develop. We therefore tried to distinguish between subduction zones with marginal basins and back-arc spreading (Karig, 1971a, 1971b, 1972) and those without. Of the six ridges in the latter category (Alaska Seamounts, Cocos, Carnegie, Nazca, the Tehuantepec Ridges, and the Juan Fernandez chain) only the Tehuantepec Ridge and Juan Fernandez chain trend into local cusps. However, there are other effects apparently associated with downgoing ridges, whether back-arc extension occurs or not. A region of reduced seismicity seems to be associated with many consumed ridges. Again, the effect could be an artifact of insufficient sampling time, but it occurs in too many areas to be generally explained in this way. True, the existence of a zone of reduced seismicity is dubious in some places (for example, Fig. 17), and in other cases it may simply be a secondary effect of reduced convergence rate or of the geometry of the cusp (Fig. 9). Further work is required to establish whether the effect is restricted to certain earthquake depth or magnitude ranges. Even if the entire subduction zone is eventually "covered" by great earthquakes and their aftershocks (Sykes, 1971; Kelleher and others, 1974), this may not be true for the more common, lesser events such as the great majority depicted in this paper. Also, the statistics for smaller events are better, insofar as aseismic gaps are less likely to be artifacts of the record interval. The reduced seismicity might reflect (1) a thinner, hotter plate below the ridge; (2) a greater tendency to deform by creep rather than brittle fracture; (3) the extra basalt magmas derived from the subducted ridge and intrusives at depth, which would reduce the strength of the consuming and consumed plates; and (4) the tendency for the consumed plate to become detached along the ridge, the latter being a probable line of weakness. Point (1) would be expected if the ridges were made by hot spots and are younger than the surrounding crust. The relative age of ridge and crust is generally only poorly known, however. There is no general correlation between the presence or absence of Quaternary andesite volcanoes and subducted ridges (Fig. 1), a surprising observation in light of the thickened basalt crust available. We, therefore, do not believe explanation (3) is important. Point (4) is supported by the case of the projected Louisville Ridge, which seems to mark the trace of a major seismic discontinuity (Fig. 9) in the consumed slab. But this could also mean the fracture zone associated with the ridge has been reactivated as the slab went down. The Nazca, Cocos, Carnegie, and Juan Fernandez Ridges are other cases where the downgoing ridge bounds dissimilar seismic, volcanic, or other tectonic provinces. Thus, many andesite volcanic chains or particular magmatic or tectonic provinces stop or start near the intersection points (Sillitoe, 1974) (Fig. 1). We speculate that the downgoing ridges are transverse lines of weakness that tend to separate the slabs into individual tongues of slightly different behavior. Volcanoes develop along one tongue or another depending on circumstances presently unknown; the point we make is only that each tongue behaves for volcanic or seismic purposes as a unit. Stauder (1973) and Sillitoe (1974) have similarly concluded that the slabs are segmented into tongues.

The mechanisms we propose (Figs. 3, 13, 14) are scarcely more than geometrical; the usual admonition for further work is particularly apt here. In many areas, the time of cusp formation is unknown, thus precluding one of the tests of the proposed mechanism. Often the cusp formation time is not identical to the age of oldest rocks in the arc. The timing of back-arc spreading (Karig, 1971a, 1971b) and paleomagnetic evidence for arc rotation are two "handles" on this timing problem. Detailed geological studies in the intersection area are also needed. Are there scraped-off seamounts like, perhaps, the Bonin Islands (Fig. 4)? Can the time of first collision be identified? Are the andesite volcanics, if any, different in composition where a ridge has gone down? Finally, seismicity should be examined in the intersection area to look for thinning or detachment of the slab, distinctive magnitude-frequency relations, or other peculiarities that may shed light on the physical processes. Another decade of seismic observation should reveal whether, for example, the Alaska (Fig. 10) and west Aleutian (Fig. 9) aseismic regions are artifacts. Finally, what is the origin of the cusps and arcs of suture zones like the Alpine-Himalayan belt and older ones? Are they relict island-arc traces, their cusps formed by aseismic ridges on the floors of vanished oceans?

Acknowledgments

We thank H. C. Eppert, Jr., T. L. Holcombe, H. Kanamori, and K. E. Louden for review and B. Wells, R. Edman, N. Hunt, and R. Boerckel for assistance. This work was partially supported by the Office of Naval Research.

Selected Bibliography

Andrews, J. D., 1971, Gravitational subduction of a western Pacific crustal plate: Nature Phys. Sci., v. 233, p. 81–82.

Anonymous, 1970a, Tectonic map of the Pacific segment of the Earth: Moscow, Geol. Inst. and Inst. Oceanology of Acad. Sciences USSR.

Anonymous, 1970b, Bathymetric atlas of the northwestern Pacific Ocean: Prepared by Scripps Institution of Oceanography under the direction of T. E. Chase and H. W. Menard, H. O. Pub. No. 1301-S, U.S. Naval Oceanog. Office.

Anonymous, 1971a, Bathymetric atlas of the northeastern Pacific Ocean: Prepared by Scripps Institution of Oceanography under the direction of T. E. Chase and H. W. Menard, H. O. Pub. No. 1303, U.S. Naval Oceanog. Office.

Anonymous, 1971b, Bathymetric atlas of the northcentral Pacific Ocean: Prepared by Scripps Institution of Oceanography under the direction of T. E. Chase and H. W. Menard, H. O. Pub. No. 1302-S, U.S. Naval Oceanog. Office.

Anonymous, 1972a, The World Series 1142, 1st ed.: Washington, D.C., Department of Defense.

Anonymous, 1972b, Preliminary determination of earthquake epicenters: Natl. Earthquake Inf. Center, NOAA, Washington, D.C. (tape).

Atwater, T., 1970, Implications of plate tectonics for the Cenozoic tectonic evolution of western North America: Geol. Soc. America Bull., v. 81, p. 3513–3536.

Baker, P. E., 1968, Comparative volcanology and petrology of the Atlantic island arcs: Bull. Volcanol., v. 32, p. 189–206.

Barker, P. F., 1972, A spreading centre in the east Scotia Sea: Earth and Planetary Sci. Letters, v. 15, p. 123–132.

Barker, P. F., and Griffiths, D. H., 1972, The evolution of the Scotia Ridge and Scotia Sea: Royal Soc. London Philos. Trans., Ser. A., v. 271, p. 151–183.

Ben-Avraham, Z., Segawa, J., and Bowin, C., 1972, An extinct spreading center in the Philippine Sea: Nature, v. 240, p. 453–455.

Bott, H.H.P., 1971, The interior of the earth: London, Edwards Arnold, p. 316.

Bowin, C., 1975, The geology of Hispaniola, *in* Nairn, A.E.M., and Stehli, F. G., eds., The ocean basin and margins, Vol. 3: New York, Plenum Press (in press).

Bracey, D. R., 1975, Reconnaissance geophysical survey of the Caroline Basin: Geol Soc. America Bull., v. 86, p. 775–784.

Bracey, D. R., and Andrews, J. E., 1974, Western Caroline Ridge: Relic island arc?: Marine Geophys. Research, v. 2, p. 111–125.
Bracey, D. R., and Vogt, P. R., 1970, Plate tectonics in the Hispaniola area: Geol. Soc. America Bull., v. 81, p. 2855–2860.
Bunce, E. T., Phillips, J. D., Chase, R. L., and Bowin, C. O., 1970, The Lesser Antilles arc and eastern margin of the Caribbean Sea, *in* Maxwell, A. E., ed., The sea, Vol. 4: New York, Interscience Pubs., Inc., p. 359–385.
Burk, C. A., 1965, Geology of the Alaska Peninsula—Island arc and continental margin (Pt. 1): Geol. Soc. America Mem. 99, 250 p.
——1972, Uplifted eugeosynclines and continental margins, *in* Shagam, R., and others, eds., Studies in Earth and space sciences (Hess Volume): Geol. Soc. America Mem. 132, p. 75–85.
Casertano, L., 1963, General characteristics of active Andean volcanoes and a summary of their activities during recent centuries: Seismol. Soc. America Bull., v. 53, p. 1415–1433.
Chase, C. G., 1971, Tectonic history of the Fiji Plateau: Geol. Soc. America Bull., v. 82, p. 3087–3110.
Clague, D. A., and Jarrard, R. D., 1973, Tertiary Pacific plate motion deduced from the Hawaiian-Emperor chain: Geol. Soc. America Bull., v. 84, p. 1135–1154.
Dalziel, I.W.D., and Elliot, D. H., 1973, The Scotia arc and Antarctic margin, *in* Nairn, A.E.M., and Stehli, F. G., eds., The ocean basins and margins, Vol. 1: New York, Plenum Press, p. 171–246.
Dalziel, I.W.D., deWit, M. J., and Palmer, K. F., 1974, Fossil marginal basin in the southern Andes: Nature, v. 250, p. 291–294.
Den, N., Ludwig, W. J., Murauchi, S., Ewing, M., Holta, H., Asanuma, T., Yoshi, T., Kubotera, A., and Hagiwana, K., 1971, Sediments and structure of the Eauripik Rise: Jour. Geophys. Research, v. 76, p. 4711–4723.
Dengo, G., 1973, Estructura geologica, historía tectonica y morfologia de América Central: Centro Regional de Ayuda Tecnica, Mexico, 52 p.
Dewey, J. F., and Bird, J. M., 1970, Mountain belts and the new global tectonics: Jour. Geophys. Research, v. 75, p. 2615–2647.
Dickinson, W. R., 1973, Widths of modern arc-trench gaps proportional to past duration of igneous activity in associated magmatic arcs: Jour. Geophys. Research, v. 78, p. 3376–3389.
Dietz, R. S., Holden, J. C., and Sproll, W. P., 1970, Geotectonic evolution and subsidence of Bahama Platform: Geol. Soc. America Bull., v. 81, p. 1915–1928.
Ewing, J. I., Edgar, N. T., and Antoine, J. W., 1970, Structure of the Gulf of Mexico and Caribbean Sea, *in* Maxwell, A. E., ed., The sea, Vol. 4: New York, Interscience Pubs., Inc., p. 359–385.
Ewing, J. I., Ludwig, W. J., Ewing, M., and Eittreim, S. L., 1971, Structure of the Scotia Sea and Falkland Platform: Jour. Geophys. Research, v. 76, p. 7118–7137.
Fink, L. K., Jr., 1972, Bathymetric and geologic studies of the Guadaloupe region, Lesser Antilles arc: Marine Geology, v. 12, p. 267–288.

Fisher, A. G., and others, 1971, Initial reports of the Deep Sea Drilling Project, Vol. VI: Washington, U.S. Govt. Printing Office, p. 389–561.

Fitch, T. J., 1972, Plate convergence, transcurrent faults, and internal deformation adjacent to southeast Asia and the western Pacific: Jour. Geophys. Research, v. 77, p. 4432–4460.

Frank, F. C., 1968, Curvature of island arcs: Nature, v. 220, p. 363.

Grow, J. A., and Atwater, T., 1970, Mid-Tertiary tectonic transition in the Aleutian arc: Geol. Soc. America Bull., v. 81, p. 3715–3722.

Hayes, D. E., and Ewing, M., 1970, Pacific boundary structure, *in* Maxwell, A. E., ed., The sea, Vol. 4: New York, Interscience Pubs., Inc., p. 29–72.

——1971, The Louisville Ridge—A possible extension of the Eltanin fracture zone, *in* Reid, J. L., ed., Antarctic oceanology, Vol. 1: Washington, D.C., Am. Geophys. Union, Antarctic Research Series, v. 15, p. 223–228.

Herron, E. M., 1972, Sea-floor spreading and Cenozoic history of the east-central Pacific: Geol. Soc. America Bull., v. 83, p. 1671–1692.

Hess, H. H., 1948, Major structural features of the western North Pacific: Geol. Soc. America Bull., v. 59, p. 417–446.

Hess, H. H., and Maxwell, J. C., 1953, Caribbean research project: Geol. Soc. America Bull., v. 64, p. 1–6.

Holcombe, T. L., Vogt, P. R., Matthews, J. E., and Murchison, R. R., 1974, Evidence for sea-floor spreading in the Cayman Trough: Earth and Planetary Sci. Letters, v. 20, p. 357–371.

Jacobs, J. A., Russell, R. D., and Wilson, J. T., 1974, Physics and geology: New York, McGraw-Hill, p. 501–507.

Johnson, G. L., and Lowrie, A., 1972, Cocos and Carnegie Ridges—Result of the Galapagos "Hot Spot"?: Earth and Planetary Sci. Letters, v. 14, p. 279–280.

Johnson, T., and Molnar, P., 1972, Focal mechanisms and plate tectonics of the southwest Pacific: Jour. Geophys. Research, v. 77, p. 5000–5032.

Karig, D., Ingle, J. C., Jr., Bouma, A. H., Ellis, H., Haile, N., Koizumi, I., MacGregor, I. D., Moore, J. C., Ujiie, H., Watanabe, T., White, S. M., Yasui, M., and Yi Ling, H., 1973, Origin of the west Philippine Basin: Nature, v. 246, p. 458–461.

Karig, D. E., 1970a, Kermadec arc—New Zealand tectonic confluence, N.Z.: Jour. Geology and Geophysics, v. 13, p. 21–29.

——1970b, Ridges and basins of the Tonga-Kermadec island arc system: Jour. Geophys. Research, v. 75, p. 239-254.

——1971a, Structural history of the Mariana island arc system: Geol. Soc. America Bull., v. 82, p. 323–344.

——1971b, Origin and development of marginal basins in the western Pacific: Jour. Geophys. Research, v. 76, p. 2542–2561.

——1972, Remnant arcs: Geol. Soc. America Bull., v. 83, p. 1057-1068.

——1973, Plate convergence between the Philippines and the Ryukyu Islands: Marine Geology, v. 14, p. 153–168.

Karig, D. E., and Mammerickx, J., 1972, Tectonic framework of the New Hebrides island arc: Marine Geology, v. 12, p. 187–205.

Katsumata, M., and Sykes, L. R., 1969, Seismicity and tectonics of the western Pacific: Izu-Mariana Caroline and Ryukyu-Taiwan region: Jour. Geophys. Research, v. 74, p. 5923–5948.

Kelleher, J., Savino, J., Rowlett, H., and McCann, W., 1974, Why and where great thrust earthquakes occur along island arcs: Jour. Geophys. Research, v. 79, p. 4889–4899.

Khudoley, K. M., and Meyerhoff, A. A., 1971, Paleogeography and geological history of Greater Antilles: Geol. Soc. America Mem. 129, 199 p.

Kienle, J., 1971, Gravity and magnetic measurements over Bowers Ridge and Shirshov Ridge, Bering Sea: Jour. Geophys. Research, v. 76, p. 7138–7153.

Krause, D. C., 1973, Crustal plates of the Bismarck and Solomon Seas, *in* Fraser, R., compiler, Oceanography of the South Pacific: Wellington, New Zealand National Commission for UNESCO, p. 319–320.

Kroenke, L. W., 1974, Origin of continents through development and coalescence of oceanic flood basalt plateaus: Am. Geophys. Union Trans., v. 55, p. 444.

Ladd, J., 1974, Relative motion between North and South America and Caribbean tectonics [abs.]: VII eme Conf. Geologique de Caraibes, Ministere de l'Industrie, Antilles Francaises, p. 37.

Larson, E. E., Reynolds, R. L., Ozima, M., Aoki, Y., Kinoshita, H., Zasshu, S., Kawai, N., Nakajima, T., Hirooka, K., Merrill, R., and Levi, S., 1975, Paleomagnetism of Miocene volcanic rocks of Guam and the curvature of the southern Mariana island arc: Geol. Soc. America Bull., v. 86, p. 346–350.

Larson, R. L., and Chase, C. G., 1972, Late Mesozoic evolution of the western Pacific Ocean: Geol. Soc. America Bull., v. 83, p. 3627–3644.

Laughton, A. S., Matthews, D. H., and Fisher, R. L., 1970, The structure of the Indian Ocean, *in* Maxwell, A. E., ed., The sea, Vol. 4: New York, Interscience Pubs., Inc., p. 543–586.

Le Pichon, X., and Fox, P. J., 1971, Marginal offsets, fracture zones, and the early opening of the North Atlantic: Jour. Geophys. Research, v. 76, p. 6294–6308.

Le Pichon, X., Francheteau, J., and Bonnin, J., 1973, Plate tectonics: Development in geotectonics, Vol. 6: Amsterdam, Elsevier, p. 227–229.

Louden, K. E., and Sclater, J. G., 1974, Magnetic anomalies in the west Philippine Basin: Am. Geophys. Union Trans., v. 56, p. 1187.

Luyendyk, B. P., MacDonald, K. C., and Bryan, W. B., 1973, Rifting history of the Woodlark basin in the southwest Pacific: Geol. Soc. America Bull., v. 84, p. 1125–1134.

Luyendyk, B. P., Bryan, W. B., and Jezek, P. A., 1974, Shallow structure of the New Hebrides island arc: Geol. Soc. America Bull., v. 85, p. 1287–1300.

Malfait, B. T., and Dinkelman, M. G., 1972, Circum-Caribbean tectonic and igneous activity and the evolution of the Caribbean plate: Geol. Soc. America Bull., v. 83, p. 251–272.

Mammerickx, J., Chase, T. E., Smith, S. M., and Taylor, I. L., 1971, Bathymetry of the South Pacific: Scripps Inst. Oceanography, La Jolla, Calif., sheet 12.

Matthews, J. E., and Holcombe, T. L., 1974, Possible Caribbean underthrusting of the Greater Antilles along the Muertos Trough [abs.]: VII eme Conf. Geologique de Caraibes, Ministere de l'Industrie, Antilles Francaises, p. 44.

Maynard, G. L., Sutton, G. H., Hussong, D. M., and Kroenke, L. W., 1974, Seismic wide-angle reflection and refraction investigation of the sediments on the Ontong Java Plateau: Am. Geophys. Union Trans., v. 54, p. 378.

McKenzie, D., and Sclater, J. G., 1971, The evolution of the Indian Ocean since the Late Cretaceous: Royal Astron. Soc. Geophys. Jour., v. 25, p. 437–528.

Menard, H. W., 1966, Sea floor relief and mantle convection, *in* Physics and chemistry of the Earth, Vol. 6: Oxford, Pergamon Press, p. 315–364.

Menard, H. W., and Chase, T. E., 1970, Fracture zones, *in* Maxwell, A. E., ed., The sea, Vol. 4: New York, Interscience Pubs., Inc., p. 421–443.

Menard, H. W., and Fisher, R. L., 1958, Clipperton fracture zone in the northeastern equatorial Pacific: Jour. Geology, v. 66, p. 239–253.

Meyerhoff, A. A., and Meyerhoff, H. A., 1972, Continental drift, IV: The Caribbean "plate": Jour. Geology, v. 80, p. 34–60.

Milson, J. S., 1970, Woodlark basin, a minor center of sea-floor spreading in Melanesia: Jour. Geophys. Research, v. 75, p. 7335–7339.

Mitronovas, W., and Isacks, B., 1971, Seismic velocity anomalies in the upper mantle beneath the Tonga-Kermadec island arc: Jour. Geophys. Research, v. 76, p. 7154–7180.

Moberly, R., 1972, Origin of lithosphere behind island arcs, with reference to the western Pacific, *in* Shagam, R., and others, eds., Studies in Earth and space science (Hess volume): Geol. Soc. America Mem. 132, p. 35–73.

Molnar, P., and Sykes, L. R., 1969, Tectonics of the Caribbean and Middle America regions from focal mechanisms and seismicity: Geol. Soc. America Bull., v. 80, p. 1639–1684.

Moore, J. C., 1973, Cretaceous continental margin sedimentation, southwestern Alaska: Geol. Soc. America Bull., v. 84, p. 595–614.

Morgan, W. J., 1971, Convection plumes in the lower mantle: Nature, v. 230, p. 42–43.

——1972, Deep mantle convection plumes and plate motions: Am. Assoc. Petroleum Geologists Bull., v. 56, p. 203–213.

——1973, Plate motions and deep mantle convection, *in* Shagam, R., and others, eds., Studies in Earth and space sciences (Hess Volume): Geol. Soc. America Mem. 132, p. 7–22.

Ocola, L., 1966, Earthquake activity of Peru, *in* Steinhart, J. S., and Smith, T. J., eds., Geophysical Monograph 10: Am. Geophys. Union, p. 509–528.

Peter, G., Erickson, B., and Grim, P. J., 1970, Magnetic structure of the Aleutian Trench and northeast Pacific basin, *in* Maxwell, A. E., ed., The sea, Vol. 4: New York, Interscience Pubs., Inc., p. 191–222.

Peter, G., Schubert, C., and Westbrook, G., 1974, Caribbean Atlantic geotraverse: Geotimes, v. 19, no. 8, p. 12–19.

Pichler, H., and Zeil, W., 1969, Andesites of the Chilean Andes, *in* McBirney, A. R., ed., Proceedings of the Andesite Conference, July 1–6, 1968, Oregon, USA, International Upper Mantle Project, Sci. Report #16: Oregon Dept. Geology and Mineral Industries Bull. 65, p. 175–184.

Pitman, W. C., III, and Talwani, M., 1972, Sea-floor spreading in the North Atlantic: Geol. Soc. America Bull., v. 83, p. 619–646.

Plafker, G., and Savage, J. C., 1970, Mechanism of the Chilean earthquakes of May 21 and 22, 1960: Geol. Soc. America Bull., v. 81, p. 1001–1030.

Pratt, R. M., Rutstein, M. S., Walton, F. W., and Buschur, J. A., 1972, Extension of Alaskan structural trends beneath Bristol Bay, Bering Shelf, Alaska: Jour. Geophys. Research, v. 77, p. 4994–4999.

Scholl, D. W., and Creager, J. S., 1973, Geologic synthesis of Leg 19 (DSDP) results; far North Pacific, Aleutian Ridge, and Bering Sea, *in* Creager, J. S., Scholl, D. W., and others, 1973, Initial reports of the Deep Sea Drilling Project 19: Washington, D.C., U.S. Govt. Printing Office, p. 897–913.

Scholl, D. W., Greene, H. G., and Marlow, M. S., 1970, Eocene age of the Adak "Paleozoic(?)" rocks, Aleutian Islands, Alaska: Geol. Soc. America Bull., v. 81, p. 3583–3592.

Scholl, D. W., Cooper, A. K., and Marlow, M. S., 1974, Magnetic lineations in the Bering Sea marginal basin: Am. Geophys. Union Trans., v. 55, p. 232.

Scholz, C. H., and Page, R., 1970, Buckling in island arcs [abs.]: Am. Geophys. Union Trans., v. 51, p. 429.

Sclater, J. G., and Fisher, R. L., 1974, Evolution of the east central Indian Ocean, with emphasis on the tectonic setting of the Ninetyeast Ridge: Geol. Soc. America Bull., v. 85, p. 683–702.

Sclater, J. G., and Menard, H. W., 1967, Topography and heat flow of the Fiji Plateau: Nature, v. 216, p. 991–993.

Sclater, J. G., Anderson, R. N., and Bell., M. L., 1971, Elevation of ridges and evolution of the central eastern Pacific: Jour. Geophys. Research, v. 76, p. 7888–7915.

Sclater, J. G., Hawkins, J. W., Mammerickx, J., and Chase, C. G., 1972, Crustal extension between the Tonga and Lau Ridges: Petrologic and geophysical evidence: Geol. Soc. America Bull., v. 83, p. 505–518.

Shakal, A. F., and Willis, D. E., 1972, Estimated earthquake probabilities in the north circum-Pacific area: Seismol. Soc. America Bull., v. 62, p. 1397–1410.

Sheridan, R. E., Drake, C. L., Nafe, J. E., and Hennion, J., 1966, Seismic refraction study of continental margin east of Florida: Am. Assoc. Petroleum Geologists Bull., v. 50, p. 1972–1991.

Shor, G. G., Jr., and Fisher, R. L., 1961, Middle America Trench: Seismic refraction studies: Geol. Soc. America Bull., v. 72, p. 721.

Sillitoe, R. H., 1974, Tectonic segmentation of the Andes: Implications for magmatism and metallogeny: Nature, v. 250, p. 542–545.

Sleep, N. H., and Toksöz, M. N., 1971, Evolution of marginal basins: Nature, v. 233, p. 548–550.

Solomon, S., and Biehler, S., 1969, Crustal structure from gravity anomalies in the southwest Pacific: Jour. Geophys. Research, v. 74, p. 6696–6701.

Stauder, W., 1973, Mechanism and spatial distribution of Chilean earthquakes with relation to subduction of the oceanic plate: Jour. Geophys. Research, v. 78, p. 5033–5061.

Sykes, L. R., 1966, The seismicity and deep structure of island arcs: Jour. Geophys. Research, v. 71, p. 2981–3006.

——1971, Aftershock zones of great earthquakes, seismicity gaps, and earthquake prediction of Alaska and the Aleutians: Jour. Geophys. Research, v. 76, p. 8021–8041.

ykes, L. R., Oliver, J., and Isacks, B., 1970, Earthquakes and tectonics, *in* Maxwell, A. E., eds., The sea, Vol. 4: New York, Interscience Pubs., Inc., p. 353–420.

alwani, M., 1960, Gravity anomalies in the Bahamas and their interpretation [Ph.D. thesis]: New York, Columbia Univ., p. 1–89.

ruchan, M., and Larson, R. L., 1973, Tectonic lineaments on the Cocos plate: Earth and Planetary Sci. Letters, v. 17, p. 426–432.

urner, D. L., Forbes, R. B., and Naeser, C. W., 1973, Radiometric ages of Kodiak Seamount and Giacomini Guyot, Gulf of Alaska: Implications for circum-Pacific tectonics: Science, v. 182, p. 579–581.

Jyeda, S., and Ben-Avraham, Z., 1972, Origin and development of the Philippine Sea: Nature, v. 240, p. 176–178.

Jyeda, S., and Miyashiro, A., 1974, Plate tectonics and the Japanese Islands: A synthesis: Geol. Soc. America Bull., v. 85, p. 1159–1170.

an Andel, Tj. H., Heath, G. R., Malfait, B. T., Heinrichs, D. R., and Ewing, J. I., 1971, Tectonics of the Panama Basin, eastern equatorial Pacific: Geol. Soc. America Bull., v. 82, p. 1489–1508.

/ogt, P. R., 1973a, Subduction and aseismic ridges: Nature, v. 241, p. 189–191.

——1973b, Early events in the opening of the North Atlantic, *in* Tarling, D. H., and Runcorn, S. K., eds., Implications of continental drift to the Earth sciences: London, Academic Press, p. 693–712.

/ogt, P. R., and Bracey, D. R., 1973, Geophysical investigations of the Caroline Basins, *in* Fraser, R., eds., Oceanography of the South Pacific: Wellington, New Zealand Natl. Commission for UNESCO, p. 319–320.

Vilson, J. T., 1965, Evidence from ocean islands suggesting movement in the Earth: Royal Soc. London Philos. Trans., Ser. A, v. 258, p. 145–165.

Vinterer, E. L., and others, 1971, Initial reports of the Deep Sea Drilling Project, Vol. VII: Washington, D.C., U.S. Govt. Printing Office, p. 49, 472.

ANUSCRIPT RECEIVED BY THE SOCIETY OCTOBER 18, 1974
EVISED MANUSCRIPT RECEIVED MAY 19, 1975
ANUSCRIPT ACCEPTED JULY 15, 1975